T0062633

The ARTEMIS Mission

Christopher Russell • Vassilis Angelopoulos
Editors

The ARTEMIS Mission

Previously published in *Space Science Reviews* Volume 165,
Issues 1–4, 2011

Editors
Christopher Russell
University of California Los Angeles
Los Angeles, CA, USA

Vassilis Angelopoulos
University of California Los Angeles
Los Angeles, CA, USA

ISBN 978-1-4939-4854-3
ISBN 978-1-4614-9554-3 (eBook)
DOI 10.1007/978-1-4614-9554-3
Springer New York Heidelberg Dordrecht London

Cover illustration: An artist's concept of ARTEMIS-P1 and P2 in lunar orbit.
Credit: NASA/Conceptual Image Lab.

Printed on acid-free paper

Springer is part of Springer Science+Business Media (www.springer.com)

Contents

DOI 10.1007/978-1-4614-9554-3_1
Reprinted from *Space Science Reviews* Journal, DOI 10.1007/s11214-012-9875-3

Foreword

Acceleration Reconnection Turbulence and Electrodynamics of Moon's Interaction with the Sun (ARTEMIS) mission

C.T. Russell · V. Angelopoulos

Published online: 22 March 2012

The Moon's space environment, tenuous neutral exosphere, wake, surface and interior have been studied with single spacecraft missions since the beginning of the space age: The Apollo missions extended not only the human experience to the lunar surface but also carried our extended eyes and ears to the lunar surface and into lunar orbit to study for the first time how the solar wind interacts with airless bodies and, using magnetic induction, to probe the interior core properties. Since then, the US Clementine, Lunar Prospector and Lunar Reconnaissance Orbiter missions, ESA's SMART-1, the Japanese Kaguya, the Indian Chandrayaan and the Chinese Chang'e I and II missions have revealed with greater detail properties of lunar crustal magnetism and surface composition, while analysis of Apollo samples have rewritten textbooks regarding the formation of the Earth-Moon system and Solar System evolution. It has become apparent recently that to make further progress in understanding the complex interaction of the Moon's surface with its space environment, as well as to better probe the Moon's interior with natural electromagnetic signals from above, single point measurements are inadequate. The solar wind buffeting the surface never stays constant: it varies on the time-scales that correspond to deep electromagnetic sounding and on spatial scales comparable to several lunar radii, rendering distant spacecraft (e.g., ones at the Earth-Sun Lagrange point) ineffective monitors of the local drivers.

At a distance of 55–65 Earth radii, the Moon spends most of its time upstream of Earth's bow shock, and is thus exposed to either the pristine solar wind or to showers of particles emitted by the solar wind-magnetosphere interaction. Once per lunar month the Moon traverses Earth's magnetotail at a woefully undersampled region where magnetic reconnection, plasma turbulence and particle acceleration are commonplace and have far-reaching consequences for magnetospheric plasma circulation. Lunar orbits, therefore, are ideal platforms for observation and study of the pristine solar wind that impacts Earth, distant effects at shocks, and the dominant method of energy release in the magnetosphere, namely distant magnetotail reconnection. Typical studies of these phenomena have, again, been con-

C.T. Russell (✉) · V. Angelopoulos
University of California, Los Angeles, Los Angeles, CA, USA
e-mail: ctrussel@igpp.ucla.edu

ducted only with single spacecraft missions. Thus, spatial extent, spatio-temporal ambiguities, three-dimensional effects and dynamical evolution could only be loosely inferred in the past but not duly explored.

Recognizing the scientific return from a two-point investigation from the Moon and of the Moon, the NASA/THEMIS team embarked upon a bold initiative to use the excess fuel capacity of two of the five spacecraft in the THEMIS constellation orbiting Earth and send them via low-thrust trajectories in orbit about the Moon. From those stable, yet scientifically optimal orbits, the two spacecraft would conduct unprecedented observations of the solar wind driver and its interaction with the lunar exosphere, crustal magnetism, lunar surface (wake), and lunar interior, as well as perform two-point observations of the solar wind, its interaction with the magnetosphere, and magnetotail reconnection and their effects (plasmoids, particle acceleration and turbulence) in the distant magnetotail. Hence the ARTEMIS mission was born out of THEMIS.

ARTEMIS has been a mission of several "firsts". It is the first mission to employ two identical coordinated spacecraft at the Moon, the first mission to conduct prolonged (9 month) observations and station-keeping at an Earth-Moon Lagrange point, the first to provide DC electric field measurements at the lunar space environment, and the first to provide comprehensive in-situ measurements in the lunar space environment from a high altitude lunar orbit.

In the compendium of papers in this volume V. Angelopoulos describes the ARTEMIS mission, and T. Sweetser and colleagues present its circuitous, mission-enabling, low thrust trajectory design. Its broad scientific objectives spanning heliophysics and planetary goals are described by D. Sibeck and colleagues, and the first results from the mission, obtained during a lunar flyby enroute to a translunar orbit injection by the ARTEMIS P1 spacecraft are presented by J. Halekas and colleagues. The latter provides an exposition of the scientific potential from the instrumentation aboard that has been operating nominally five years after launch of the ARTEMIS hardware into space. The spacecraft and instrumentation, including salient features of calibration and generic data interpretation, have been described in the THEMIS Space Science Reviews volume (Angelopoulos 2008).

We hope that the present volume will be useful to researchers in both heliophysics and planetary physics for understanding published scientific results from the mission and in order to aid their own analysis and data interpretation efforts from these publicly available datasets.

February 23, 2012

Acknowledgements We greatly appreciate the efforts of the editorial staff at the Space Science Reviews, in particular Mr. Randy Cruz for his rapid and efficient processing of the submissions, and the figures, Mr. Emmanuel Masongsong for interfacing with authors and editors, Ms. Judy Hohl for editorial assistance with the manuscripts and Ms. Marjorie Sowmendran at the University of California, Los Angeles, for interfacing with the referees, authors and publishers.

We are grateful to Dr. Ayako Matsuoka who stepped in to edit the articles for which the editors were conflicted. The editors also benefited from an excellent group of referees who helped refine the articles provided by the authors. These referees included: Dennis Byrnes, Robert Farquhar, Yoshifumi Futaana, Lonnie Hood, Shinobu Machida, Tomoko Nakagawa, Masaki N Nishino, Yoshifumi Saito.

References

V. Angelopoulos, The THEMIS Mission. Space Sci Rev **141**, 5–34 (2008). doi:10.1007/s11214-008-9336-1

DOI 10.1007/978-1-4614-9554-3_2
Reprinted from *Space Science Reviews* Journal, DOI 10.1007/s11214-010-9687-2

The ARTEMIS Mission

V. Angelopoulos

Received: 2 December 2009 / Accepted: 19 August 2010 / Published online: 3 November 2010

Abstract The Acceleration, Reconnection, Turbulence, and Electrodynamics of the Moon's Interaction with the Sun (ARTEMIS) mission is a spin-off from NASA's Medium-class Explorer (MIDEX) mission THEMIS, a five identical micro-satellite (hereafter termed "probe") constellation in high altitude Earth-orbit since 17 February 2007. By repositioning two of the five THEMIS probes (P1 and P2) in coordinated, lunar equatorial orbits, at distances of $\sim$55–65 R_E geocentric ($\sim$1.1–12 R_L selenocentric), ARTEMIS will perform the first systematic, two-point observations of the distant magnetotail, the solar wind, and the lunar space and planetary environment. The primary heliophysics science objectives of the mission are to study from such unprecedented vantage points and inter-probe separations how particles are accelerated at reconnection sites and shocks, and how turbulence develops and evolves in Earth's magnetotail and in the solar wind. Additionally, the mission will determine the structure, formation, refilling, and downstream evolution of the lunar wake and explore particle acceleration processes within it. ARTEMIS's orbits and instrumentation will also address key lunar planetary science objectives: the evolution of lunar exospheric and sputtered ions, the origin of electric fields contributing to dust charging and circulation, the structure of the lunar interior as inferred by electromagnetic sounding, and the lunar surface properties as revealed by studies of crustal magnetism. ARTEMIS is synergistic with concurrent NASA missions LRO and LADEE and the anticipated deployment of the International Lunar Network. It is expected to be a key element in the NASA Heliophysics Great Observatory and to play an important role in international plans for lunar exploration.

Keywords THEMIS · ARTEMIS · Magnetosphere · Reconnection · Solar wind · Turbulence · Lunar exosphere

1 Introduction

The "Acceleration, Reconnection, Turbulence, and Electrodynamics of the Moon's Interaction with the Sun" (ARTEMIS) mission is a two-spacecraft ("probe") complement that

V. Angelopoulos (✉)
IGPP/ESS UCLA, Los Angeles, CA 90095-1567, USA
e-mail: vassilis@ucla.edu

addresses key science questions related to both heliophysics science as observed from/at the lunar environment and the lunar exosphere, surface, and interior. The mission concept utilizes the two outermost satellites of the NASA MIDEX mission THEMIS (Angelopoulos 2008), a five identical satellite mission launched on 17 February 2007 to study the origin of the magnetospheric substorms, a fundamental space weather process (Sibeck and Angelopoulos 2008).

From distances hundreds of kilometers to 120,000 km (from the Moon and at variable inter-probe separations optimized for heliophysics science, the two ARTEMIS probes will study: (i) particle acceleration, reconnection, and turbulence in the magnetosphere and (ii) in the solar wind; and (iii) the electrodynamics of the lunar environment. With its unique, coordinated, two-point measurements, ARTEMIS will reveal the dynamics, scale size, and evolution of the distant tail, the 3-dimensional structure of solar wind shocks, and the structure, evolution and kinetic properties of the lunar wake. ARTEMIS builds on our understanding of the magnetotail and solar wind environment at lunar distances that was acquired from ISEE3 (Tsurutani and von Rosenvinge 1984), Geotail (Nishida 1994) and Wind (Acuña et al. 1995). ARTEMIS will also advance our understanding the Moon's wake going beyond the observations from Wind high altitude (10 R_L) wake crossings and the low ($\sim$100 km) altitude wake and exospheric observations by Lunar Prospector (LP, Hubbard et al. 1998; Binder 1998), Kaguya (e.g., Saito et al. 2010) and Chang'E. ARTEMIS's comprehensive plasma and fields observations over an extensive range of distances from low to high altitudes fill an observational gap in wake behavior and extend the measurement capability by including DC electric field observations and a two spacecraft complement. ARTEMIS's multi-point observations, orbits, and instrumentation are also ideally suited to advance our knowledge of several key topics raised in the 2003 National Research Council's (NRC) Decadal Survey for Solar System Exploration and several prioritized science concepts listed in the 2007 National Academy of Sciences (NAC) report, "The Scientific Context for Exploration of the Moon". With all its instruments operating flawlessly and from the achievable 100 km perigee altitude, $\sim 10^\circ$ inclination orbit, ARTEMIS could contribute greatly to our understanding of the formation and evolution of the exosphere, dust levitation by electric fields, the crustal fields and regolith properties and the interior of the Moon. By optimizing periselene to obtain low-altitude passes below 100 km and inclinations as high as 20° to reach the outskirts of the South Pole—Aitken basin, the ARTEMIS team can further optimize the science return from the mission for planetary science in its prime or extended phase.

This paper describes the ARTEMIS mission concept. Following an overview of the mission history, instrument and spacecraft capabilities, and mission phases (Sect. 1), Sect. 2 presents the scientific objectives in relation to the mission design. Section 3 discusses the aspects of mission design that enabled optimal science within the capabilities of spacecraft already in orbit. Section 4 describes the unique features of the ARTEMIS operations that were critical in achieving the heliophysics and planetary aspects of the mission. This section also provides an overview of the data processing and data dissemination system as it has evolved through the successful THEMIS mission practices and is now applied on ARTEMIS. Detailed aspects of the scientific objectives, mission design, navigation, operations, and first results will be presented in future publications.

2 Overview

ARTEMIS arose well into the THEMIS mission's Phase-C development cycle, when it was recognized that Earth shadows exceeding the spacecraft bus thermal design limits would threaten THEMIS probes TH-B (P1) and TH-C (P2) during their third tail season. This was destined to happen on March 2010, about six months after the end of the prime mission (Fig. 1(a)). Additionally, at that time the angles between the lines of apsides for P1 and P2 would be 54° and 27° away from those of P3, P4, and P5, rendering the classic five-probe conjunctions of the prime THEMIS mission design non-optimal. Preliminary studies by NASA/JPL in 2005 indicated that by placing P1 and P2 into lunar orbits (Fig. 1(b)) using a low-thrust lunar capture mission design, the risk of freezing would be avoided as the shadow durations would become small and manageable. The potential of P1 and P2 for scientific discovery could be further maximized for heliophysics science by careful optimization of the mission design to result in variable inter-probe separation vectors relative to the Sun-Moon and Sun-Earth line. This optimization was the genesis of the ARTEMIS concept. An ARTEMIS science team was formed at that point to define the scientific goals of the mission and worked on science optimization. The mission was approved by the NASA Heliophysics Senior Review panel in May 2008 (http://wind.nasa.gov/docs/Senior_Review_2008_Report_Final.pdf), and ARTEMIS operations commenced on July 20th, 2009—coinciding with the 40th anniversary of NASA's first lunar landing.

With the prime THEMIS mission successfully completed by September 2009 and with fuel margins on P1 and P2 remaining robust, the ARTEMIS implementation is proceeding as planned. Three lunar flybys in January-March of 2010 resulting in translunar injections (TLI) will place the probes in orbits near the Earth-Sun Lagrange points. These flybys are also expected to provide a first glimpse into the type of ARTEMIS lunar wake data to be expected from the nominal mission. Following a series of Earth flybys in 2010, the two probes are expected to reach Lissajous orbits (the Lagrange points of the Earth-Moon system) in October 2010 and enter into lunar orbits in April 2011. Figure 2 shows the geometry of those orbits in a coordinate system centered at the Moon, with X-axis opposite Earth, Z axis perpendicular to the Earth-Moon orbit plane, positive North, and Y axis completing the right-hand coordinate system. Very little fuel is needed to move a probe from the Lunar Lagrange point 1 (LL1) on the Earth side, to the LL2, opposite to Earth. Very little fuel is required to maintain the spacecraft from one Lagrange point to the other, resulting in semi-periodic Lissajous orbits in this coordinate system.

The ARTEMIS team has been given the go-ahead to implement a 2 year mission. The probes' radiation safety margin, robust instrumentation, and stable orbits, however, make the mission capable of providing high quality measurements of the lunar environment during the next solar cycle. Table 1 outlines the mission phases, durations, and typical orbit separations in each phase and links them to the science objectives discussed above and in Sect. 2. The mission phases are as follows: The two probes, P1 and P2, arrive at the Lissajous orbits, on opposite sides of the Moon, on September 1, 2010 (P1, near-side) and October 19, 2010 (P2, far-side), respectively. The insertion of P2 is gradual, such that useful tail and solar wind two-probe conjunctions can commence as early as September 21, 2010. The probes stay in this configuration until January 8, 2011. In the Lissajous orbits, although the probes hover ~60,000 km away from the Moon along the Earth-Moon line (on their respective sides of the Moon), they are librating along their orbit-tracks about Earth, ±60,000 km ahead of or behind the Moon. This strategy results in a variety of P1-P2 conjunctions with inter-probe separations of 60,000–120,000 km ($dR \sim$ 10–20 R_E, or 35–70 R_L) that are either along the Sun-Earth line or across it; those conjunctions can be either in the solar wind, or in the magnetotail and magnetosheath. This strategy also results in six long-range lunar wake crossings by either P1 or P2 from around 20 and 30 R_L. Figure 3(a) shows snapshots of two possible relative positions of P1 and P2 in the magnetosphere and the solar wind. Due

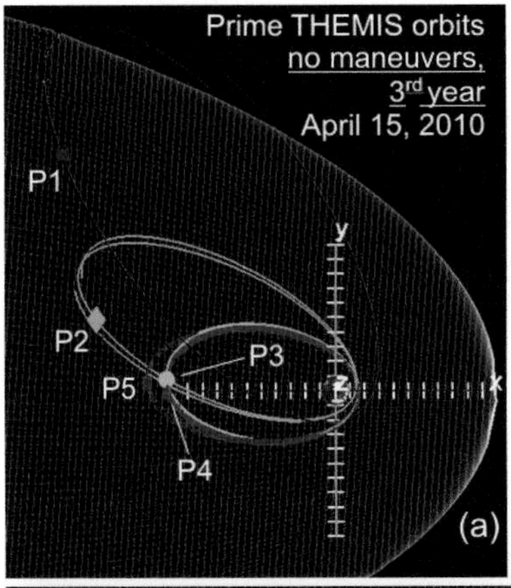

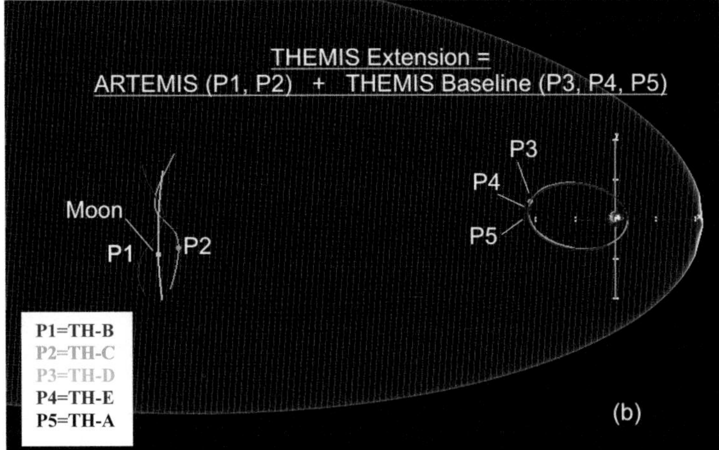

Fig. 1 (**a**) THEMIS orbits in X-Y Geocentric Solar Magnetospheric (GSM) coordinates during the 3rd year of the operations (2010) in the absence of orbitraising maneuvers for P1, P2. Differential precession of the line of apsides prevents conjunctions between probes along the Earth's magnetotail. In addition, long Earth shadows in March 2010 (not shown) would have presented a problem to mission safety. *Bottom*: The ARTEMIS mission raises the apogees of P1 and P2 such that they are captured into lunar orbits, resulting in new science from the lunar environment

to the sensitivity of the orbit profile to initial (capture) conditions and to orbit maintenance maneuvers, the exact times of those conjunctions may vary but their overall nature will remain qualitatively the same. This mission phase, which is denoted as Lunar Lagrange points 1 and 2 phase, or "LL1,2", lasts approximately 3 months.

In early January 2011, probe P1 will be commanded to leave its Lissajous orbit on the far side and enter orbit into the Lunar Lagrange point 1, or LL1, on the near-side of the Moon. At different phases of their Lissajous orbits, the two probes (P1 and P2) reside at inter-probe separation vectors of size 5–20 R_E, with longer ranges preferentially across the Earth-Moon line and shorter ranges along the Sun-Earth line. Figure 3(b) shows two snapshots of such possible configurations. Another six lunar crossings are also acquired in this phase, from distances around 10–30 R_L (most are around 15–20 R_L); those wake crossings occur typically far upstream of Earth's bow shock and are pristine, i.e., least affected by Earth

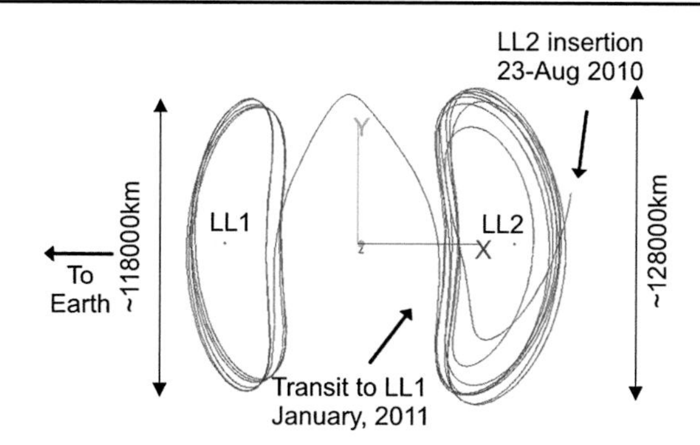

Fig. 2 ARTEMIS Lissajous orbits in the initial phase of the mission, shown here for the case of P1 (adapted from Broschart et al. 2009; also see Sweetser et al. 2010)

foreshock effects. This phase lasts another 3 months, until early April, 2011. It is denoted Lunar Lagrange point 1 phase, or "LL1" phase.

In early April 2011 P1 will be commanded to insert into lunar orbit by performing a series of periselene burn maneuvers; P2 will follow suit in late April. By May 2011 both P1 and P2 will be in stable equatorial, high-eccentricity orbits, of ~100 km × 19,000 km altitude. These are stable, 26 h period orbits with inter-probe separations 500 km–5R_E. One probe will be in a retrograde and the other in a pro-grade orbit, such that the precession rates of their line of apsides will walk relative to each other by 15–20 degrees per month; in ~2 years the lines of apsides will cover a full circle, resulting in a wide range of inter-probe vectors relative to the Sun-Earth line and lunar wake crossings from a wide range of altitudes. A subset of orbits resulting in simultaneous, two-probe crossings of the lunar wake is also possible during certain mission phases, when one of the two probes is at apoapsis along the Sun-Moon line behind the Moon (Fig. 3(c)).

The ARTEMIS spacecraft (probes) are identical. A probe in deployed configuration and the instrument field of views are shown pictorially in Fig. 4. The probes, which are spin-stabilized platforms (Harvey et al. 2008) with 3 s spin period, carry body-mounted particle instruments and tethered fields instruments. The spin axis is nominally maintained at an angle < 10 deg to the ecliptic South (unlike probes P3, 4, and 5, which have spin axes due close to ecliptic North). The probes are equipped with monopropellant hydrazine propulsion systems, capable of providing ~1.5 N of thrust at $I_{SP} \sim 210$ s near the end of mission. Maneuvers are typically side-thrusts. Axial thrusts provide, when necessary to combine with side-thrusts, a vector thrust off of the spin plane, when necessary to match precisely the specified ΔV vector. Attitude sensors include a Sun sensor, two backup rate gyros, and the science magnetometer (useful for attitude knowledge only near perigee). For ARTEMIS, away from Earth's strong field, the primary attitude sensor is thus the Sun sensor, used to derive full spin attitude information by modeling Sun motion as a function of time over a period of days to weeks. Attitude predicts from thrust and vehicle performance modeling are typically characterized well enough that the Sun sensor is used to check and re-set the absolute attitude whenever possible in-between thrust operations.

The particle instruments ESA and SST (fields of view shown in Fig. 4) measure thermal and super-thermal ions and electrons. Sun pulse information is used to sector data into 3D distribution functions over the period of one spin. The fields instruments, FGM, SCM, and EFI, measure with state-of-the-art cadence, offset stability and sensitivity the DC and AC magnetic and electric fields. Table 2 shows the main instruments and the reference in which more information about instrument characteristics can be obtained. Radiation dose margin of 2, latch-up protection circuitry and memory scrubbing have been implemented on the instruments and selectively on the spacecraft (Harvey et al. 2008). All instruments and

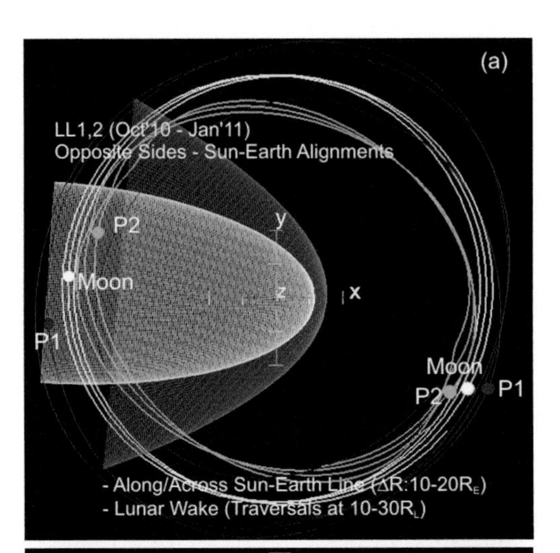

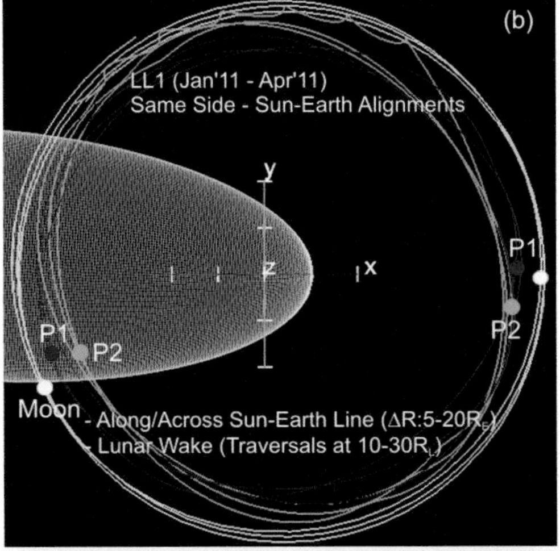

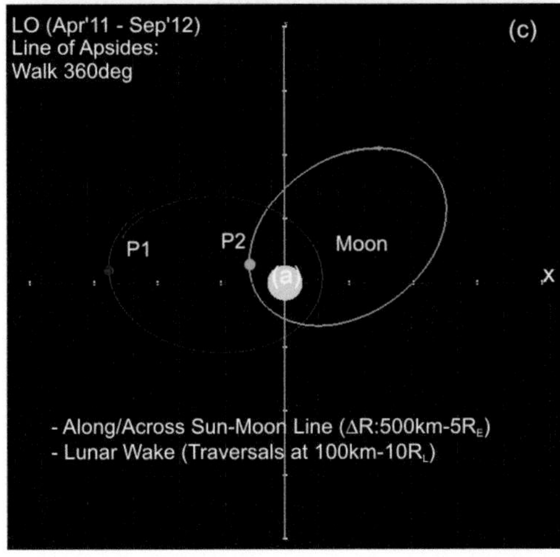

Fig. 3 (**a**) ARTEMIS orbits in the first 3 months of science operations: P1 and P2 are on Lunar Lagrange points 1 and 2, i.e., on opposite sides of the Moon, shown here in the GSM coordinate system (same as Fig. 1). Shown are two representative inter-probe separation conditions in the magnetosphere (*white mesh*) and in the solar wind (i.e., outside the shock region represented by the green mesh). (**b**) Same as in (**a**) but for the next 3 months of science operations.
(**c**) Representative ARTEMIS probe locations after lunar orbit insertion, shown in Selenocentric Solar Ecliptic coordinates, with *horizontal axis* towards the Sun (positive to the right) and *vertical axis* along the cross product of the ecliptic-normal and the Moon-Sun line (positive upwards)

Table 1 ARTEMIS orbits and mission phases relative to heliophysics and planetary objectives

Phase	Abbr	Interval	ARTEMIS: probes P1, P2	Heliophysics Objective	Planetary Objective
Translunar Injection	TLI	Oct. '09– Oct. '10	Translunar orbits to capture into LL1, LL2	Lunar flybys: Build tools, experience	Lunar flybys: Build tools, experience
P1 at LL2, P2 at LL1 ($D_L = 30$–$60\ R_L$)	LL1,2 Phase	Oct. '10– Jan. '11	$dR_{P1\text{-}P2} = 20\ R_E$ at Moon $dR_{P1\text{-}P2}$ along/across wake & Sun-Earth $dX^{GSE}_{P1\text{-}P2} \sim dY^{GSE}_{P1\text{-}P2}$ ~ 500 km–20 R_E	*In the Magnetotail* Rx, SW-magnetosphere interaction, tail turbulence *In the Solar Wind (SW)* Foreshock, shock acceleration, Rx, SW turbulence *In the Wake (SW or Tail)* Kinetics and dynamics of lunar wake in SW, sheath, tail	At solar wind (SW) wake or downstream: Pickup ions?
P1, P2 both at LL1 (each at $D_L = 30$–$60\ R_L$)	LL1 Phase	Jan. '11– Apr. '11	$dR_{P1\text{-}P2} = 5$–$20\ R_E$ at Moon $dR_{P1\text{-}P2}$ along/across wake & Sun-Earth $dX^{GSE}_{P1\text{-}P2} \sim dY^{GSE}_{P1\text{-}P2}$ ~ 500 km–20 R_E		At solar wind (SW) wake or downstream: Pickup ions?
In Lunar Orbit ($D_L = 1.1$– $12\ R_L$)	LO Phase	Apr. '11– Sep. '12	$dR_{P1\text{-}P2} = 500$ km– 20 R_L at Moon $dR_{P1\text{-}P2}$ along/across wake & Sun-Earth Periselene = ~100 km [trade TBD] Aposelene = ~19000 km Inclination = ~10 deg [trade TBD]		*In the Solar Wind (SW)* Wake/downstream: pickup ions Periselene wake: crust, core Periselene dayside: Dust *Magnetotail* Crust, mini-magnetospheres, core Periselene dayside only: Magnetotellurics, dust

Key: T = Tail; Rx = Reconnection; R_L = Lunar radii; R_E = Earth radii; D_L = Distance from Moon; $dR_{P1\text{-}P2}$ = Inter-probe separation vector; $dX^{GSE}_{P1\text{-}P2} = dR_{P1\text{-}P2}$ projection along X in Geocentric Solar Ecliptic coordinates—similar for $dY^{GSE}_{P1\text{-}P2}$, $dZ^{GSE}_{P1\text{-}P2}$

spacecraft are operating flawlessly, with no signs of performance degradation. Since the thermal and radiation design have been optimized for the worst-case environment, which is the one experienced by the inner THEMIS probes (on one-day period orbits, at 12 R_E apogee), the outer probes have seen significantly less cumulative thermal cycling and radiation (by approximately a factor of 2–4) relative to their design limit. Additionally, the outer probes were, by selection, the ones with more robust communication and power systems behavior. By virtue of the probes' stable lunar orbits, the relatively benign radiation environment at the Moon and the ARTEMIS operations (Sect. 3), it is expected that the ARTEMIS lifetime will be a good fraction of the upcoming solar cycle.

3 Science Objectives

This section is an outline of the ARTEMIS mission's key scientific objectives as they relate to mission requirements. Although ARTEMIS was designed to address heliophysics science objectives, moderate mission redesign enables it to optimize its observation strategy to address planetary objectives, as well.

3.1 Heliophysics Science Objectives

Figure 5 shows the three regimes to be visited by the ARTEMIS probes once the science operations phase has commenced, i.e., once both ARTEMIS probes are at the lunar environ-

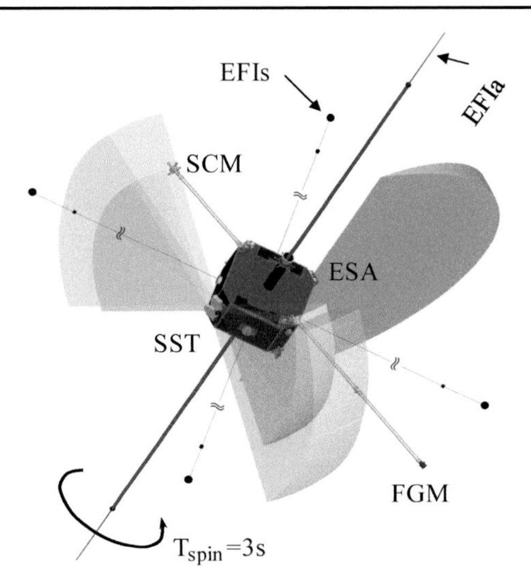

Fig. 4 Pictorial view of ARTEMIS probes and instrumentation. The spin-stabilized probes have a spin axis due ecliptic south, and a nominal spin period of 3 s. The fluxgate magnetometer (FGM) and Search Coil magnetometer (SCM) are on carbon epoxy, rigid, deployable booms of ~2 m and ~1 m length, respectively. The axial electric field (EFIa) whip sensors are at the tips of 3.5 m stacker booms and the spin plane electric field (EFIs) spherical sensors are on 20 and 25 m wire booms (truncated, not to scale in this picture). Ion and electron fields of view for the Solid State Telescope (SST) instrument are shown in different colors. The electrostatic analyzer (ESA) detector points radially outward and measures ions and electrons with identical fields of view

ment. ARTEMIS will address questions related to acceleration, reconnection, and turbulence in both regions visited by the Moon once per 28 days in its orbit about Earth: (i) the magnetosphere and (ii) the solar wind. Additionally, ARTEMIS will traverse the lunar wake routinely with one or both probes and will address questions related to (iii) the wake formation, refilling, structure, and evolution, as well as kinetic aspects of particle acceleration a the wake.

In the magnetosphere, brief passes by previous spacecraft, such as ISEE-3, Geotail, Galileo, and Wind, have demonstrated that the distant magnetotail at 55–65 R_E hosts a variety of fundamental plasma physics phenomena: quasi-steady reconnection resulting in heated plasma jets, beams of energized particles, twisted and/or unusually cold and dense plasma sheets, and turbulence. The distant reconnection line is thought to reside at 55–65 R_E from Earth, at times, making the lunar orbit particularly interesting for studies of global magnetotail circulation. The fundamental processes occurring there are common to other planetary and astrophysical systems. Additionally, the magnetotail at lunar distances is an ideal place to study the integrated output from the near-Earth processing of stored solar wind energy in the form of heated/accelerated flows and plasmoids. ARTEMIS will study these phenomena for the first time both comprehensively and systematically from the unique perspective afforded by its two identical probes. In the magnetosphere, ARTEMIS will address:

- How are particles accelerated up to hundreds of keV? Using simultaneous measurements in the lobe or mantle and in the plasma sheet, ARTEMIS will determine the mechanism of particle heating in the distant tail. The first-ever simultaneous measurements of energy inflow and particle heating will distinguish between competing particle acceleration

Table 2 ARTEMIS instruments and their capability. Survey data collection ensures plasma moments and spin fits of DC electric and magnetic fields, and nominal frequency spectra are transmitted throughout all orbits, whereas particle burst and embedded wave burst spectra ensure the highest cadence and spectral resolution fields and particles data during select intervals. Like on THEMIS, ARTEMIS bursts are selected using on-board triggers aimed at instances of high activity and include pre-burst buffers, or time-based triggers (e.g. periselenes)

Instrument	Specs	Reference
FGM: Fluxgate Magnetometer	DC magnetic field Sampling rate & resolution: DC-128 Samples/s & 3 pT Offset stability <0.2 nT/12 hr	Auster et al. (2008)
SCM: SearchCoil Magnetometer	AC Magnetic field Frequency: 1 Hz–4 kHz	Roux et al. (2008) Le Contel et al. (2008)
EFI: Electric Field Instrument	3D Electric field Frequency: DC—8 kHz	Bonnell et al. (2008) Cully et al. (2008)
ESA: Electrostatic Analyzer	Ions: 5 eV—25 keV; electrons: 5 eV–30 keV nominal g-factor/anode: ions: 0.875×10^{-3} cm^2 str electrons: 0.313×10^{-3} cm^2 str Nominal anode size: 11.25×22.5 deg minimum (solar wind ions): 5.625×5.625 deg	McFadden et al. (2008a) McFadden et al. (2008b)
SST: Solid State Telescope	Total ions: 25 keV–6 MeV Electrons: 25 keV–1 MeV	Angelopoulos (2008) for mounting and fields of view

mechanisms that have been proposed based on simulations and will determine the maximum energy obtainable under a variety of external conditions.

- What are the nature and effects of reconnection? In the absence of multipoint measurements, even the most basic characteristics of fast flows and plasmoid evolution in the tail remain poorly understood. Understanding these phenomena is important for determining how the distant tail reconnection process affects global flux and energy circulation, as well as the amount and extent of particle energization in the near-Earth environment. Radial separations of 1–10 R_E parallel to the Sun-Earth line will enable the two ARTEMIS probes and allied near-Earth spacecraft to track the evolution of high speed flows and plasmoids. Azimuthal probe separations will enable ARTEMIS to determine the cross-tail extent, orientation, shape, and topology of plasmoids. ARTEMIS will thus determine the characteristics and effects of reconnection in the distant magnetotail, from structural, magneto-hydrodynamic scales down to ion gyroradius and ion inertial length scales.
- What are the drivers and effects of turbulence? Turbulent dissipation is an effective mechanism for heating fluids and transferring mass, momentum, and energy. Characterizing the nature of these fluctuations and determining their origin and dissipation are therefore important for global circulation. Unlike the solar wind, for which time-series of near-constant velocity data can be interpreted as spatial fluctuations, tail flows are unsteady and the above simplification, enabling single spacecraft measurements of the turbulent flows, does not apply. To determine the drivers and effects of turbulence, the spatial and temporal variations of plasma and magnetic field measurements over a wide range of solar wind conditions and scale lengths must be measured. ARTEMIS's two-point measure-

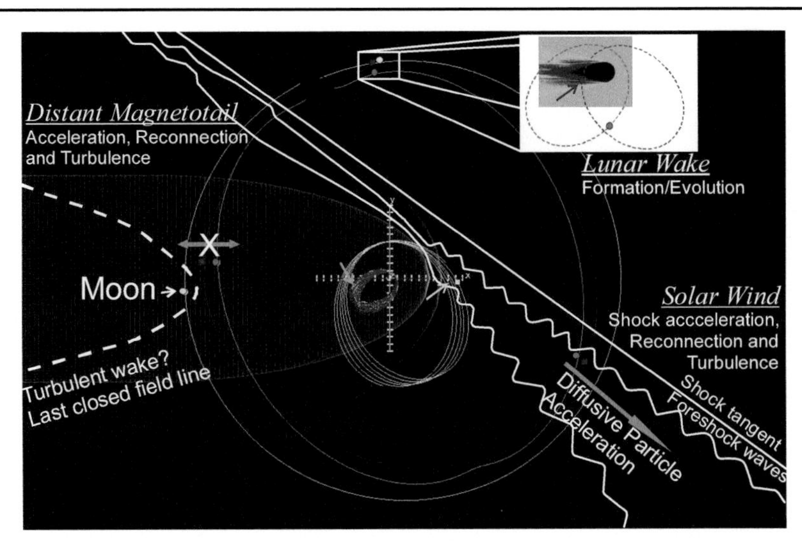

Fig. 5 ARTEMIS depicted by science region. Every 28 days probes P1 and P2 will traverse the magnetosphere, the solar wind and (multiple times) the lunar wake, addressing key questions in Heliophysics

ments at separations of a few hundred kilometers to several R_E in directions transverse to the Sun-Earth line, and in conjunction with upstream solar wind monitors, will pinpoint the origin of, establish the external conditions for, and characterize the nature of magnetotail turbulence.

In the solar wind, where ARTEMIS will spend more than 80% of its time, it will utilize its two-point measurement capability to address long-standing questions concerning the physics of solar wind particle acceleration in collisionless shocks and turbulence. Specifically, in the solar wind, ARTEMIS will address:

- How are particles accelerated at shocks? At interplanetary shocks, shock undulations are expected to host preliminary acceleration sites for solar energetic particle events and provide the seed population for particle acceleration for 10 to 100 MeV energies. Multiple spacecraft at appropriate scales are required to properly identify and study this 2D or even 3D phenomenon. Moreover, Earth's bow shock and foreshock are also excellent locations for studying the fundamental processes of particle acceleration. At lunar distances, where particles were first observed by the Apollo sub-satellites to have been diffusively accelerated at the bow shock, the acceleration process continues at rates that depend on spacecraft depth and distance to the point of tangency, as well as on upstream conditions. ARTEMIS orbits will sample the foreshock at various distances from the tangent line and at various solar wind conditions. ARTEMIS's direct inter-probe comparisons of upstream fluxes will provide a wealth of new information regarding the e-folding lengths of the diffusive acceleration process over key distances (0.1 to 20 R_E). ARTEMIS will accurately characterize the properties of interplanetary shock acceleration and diffusive particle acceleration at the Earth's bow shock and foreshock.
- What are the nature and extent of low-shear reconnection? Recent observations of reconnection "exhaust" regions have led to the identification of reconnection lines extending hundreds of Earth radii in the solar wind (Phan et al. 2006). Comprehensive examination of this phenomenon, and in particular the low-shear magnetic reconnection case, which

could be ubiquitous amongst stellar wind plasmas, is still lagging due to the scarcity of simultaneous high time resolution measurements on multiple nearby solar wind monitors. ARTEMIS's two probe high cadence plasma measurements, both alone and combined with other solar wind monitors, will enable fundamental studies of the most common, low-shear reconnection in the solar wind over scales ranging from tens to hundreds of R_E.
- What are the properties of the inertial range of turbulence? The solar wind is an excellent laboratory for the study of turbulence. Understanding the properties of the inertial range is important for modeling solar wind evolution through the heliosphere and for providing constraints on kinetic theories of energy cascade and dissipation in space plasmas in general. The crucial range, of the turbulent energy cascade at 1–20 R_E, however, has not been studied due to lack of appropriate satellite conjunctions. ARTEMIS's two-point measurements in the solar wind will fill an important gap in the study of the properties of solar wind turbulent cascade in the inertial regime and in determining how critical turbulence scale lengths vary under different solar wind conditions.

At the lunar wake, the interaction region between the solar wind and the Moon, ARTEMIS will have a unique opportunity to understand a wealth of basic physics phenomena pertaining to plasma expansion into a vacuum, applicable to many other astrophysical plasmas (voids in tori around Earth, Jupiter, and Saturn, the International space station, and Hubble). There ARTEMIS will answer:

- What are the three-dimensional structure and downstream extent of the lunar wake? Single-spacecraft wake observations cannot discern wake asymmetries arising from solar wind conditions or from crustal-solar wind interactions; this is due to either lack of pristine, nearby solar wind information or to lack of multipoint measurements. ARTEMIS's two probes will resolve spatio-temporal ambiguities at the wake. The well-instrumented probes will define the wake's extent and structure as function of downtail distance and characterize wake asymmetries.
- What are the plasma acceleration processes and energetics in and around the wake? ARTEMIS's comprehensive suite of field and plasma instruments will make possible a detailed study of the plasma physics occurring within the lunar wake that leads to acceleration and energization. The study will include the first DC electric field observations ever made in that region, direct observations of non-neutral plasma effects near the wake boundary, the extent of secondary electron beams, and their interaction with plasma refilling of the wake from the flanks.
- How do wake formation and refilling vary with solar wind and magnetospheric conditions? The wake structure varies in response to external drivers. Statistical studies provide tantalizing hints on how the wake responds to changing or transient solar wind conditions, but incomplete instrumentation and orbital coverage have limited our knowledge of this response. ARTEMIS will provide an unprecedented wealth of routine observations of the wake under a variety of solar wind conditions.

3.2 Planetary Science Objectives

The ARTEMIS team realized early on that significant benefits to planetary science could accrue from the two lunar probes with further, albeit small, orbit and instrument optimizations. ARTEMIS can address fundamental problems at the forefront of planetary science at the Moon (Fig. 6): sources and transport of exospheric and sputtered species; charging and circulation of dust by electric fields; structure and composition of the lunar interior by electromagnetic (EM) sounding; and surface properties and planetary history, as evidenced in

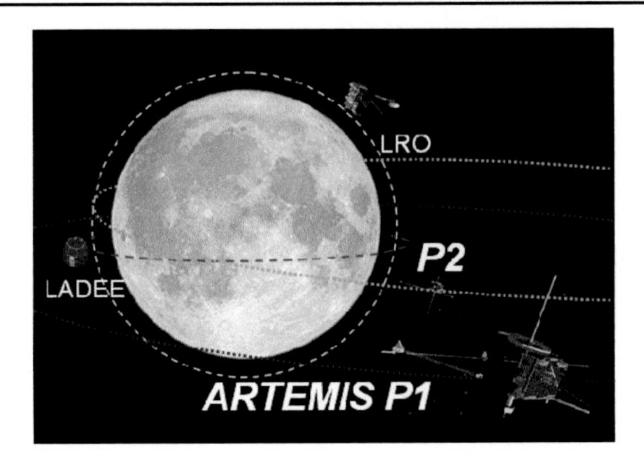

Fig. 6 ARTEMIS will study with two identical, cross-calibrated spacecraft lunar exospheric ions and dust, crustal magnetism, and the lunar interior. One probe will measure the pristine solar wind driver, while the other will study the lunar environment's response. ARTEMIS extends the SELENE/Kaguya results into the next decade, providing synergy with LRO, LADEE, and the International Lunar Network

crustal magnetism. Additionally, ARTEMIS's goals and instrumentation complement LRO's extended phase measurements of the lunar exosphere and of the lunar radiation environment by providing high fidelity local solar wind data. ARTEMIS's electric field and plasma data also support LADEE's prime goal of understanding exospheric neutral particle and dust particle generation and transport and will be in place for the deployment of the International Lunar Network (ILN), providing much needed solar wind information to ILN's studies of lunar deep electromagnetic sounding from the lunar surface.

Exosphere and sputtering. From lunar orbit, ARTEMIS will use its charged particle measurements obtained by the ESA and SST instruments as an extremely sensitive detection of ion species produced at the surface or in the exosphere and accelerated by solar wind electric fields. Newly created ions, produced by surface sputtering or ionization of exospheric gases, are generated at relatively low energies (0.01–10 eV), but immediately feel the effect of solar wind magnetic and electric fields (which ARTEMIS will also determine). Ions are then accelerated in cycloidal trajectories (i.e., "picked up"), as demonstrated by the Kaguya spacecraft (Nishino et al. 2009). Pickup ions have well-defined orbits, energy and direction as function of initial gyrophase; modeling of the observed fluxes can differentiate between surface and exospheric sources (Hartle and Killen 2006).

Lunar dust. The lunar surface electric field has been shown to respond closely to solar and magnetospheric plasma and energetic particles (Halekas et al. 2007) and also to vary with inclination with respect to the Sun. Electron reflectometry techniques have been used on Lunar Prospector (LP) to measure the potential drop between LP and the surface potential (Halekas et al. 2008a, 2009). The plasma and fields instrumentation on ARTEMIS is far more comprehensive than that flown on previous missions, enabling significant progress in our understanding of the origin and dynamics of lunar electric fields: LP measurements lacked direct knowledge of the spacecraft potential and ion measurements of any small positive potential. Although the LP potential has been modeled, ARTEMIS will be capable of actually measuring the spacecraft potential directly because it is equipped with electric field and ion analyzer instruments. Thus, ARTEMIS will go beyond the reflectometry measurements of LP with regards and measure a wide range of both positive and negative potentials.

Electromagnetic sounding. EM sounding exploits the fact that eddy currents are generated when a conductor is exposed to a changing external magnetic field. The eddy currents generate their own magnetic field, the induction field, which is readily measured by ground or space instruments. The depth to which a signal can penetrate depends on its frequency and the conductivity of the probed material. By using multiple frequencies, electromagnetic sounding has been used to probe the Earth's upper mantle (see Parkinson 1983; and references therein) and the deep lunar mantle, placing limits of ~500 km on the radius of the lunar core (Dyal et al. 1974; Russell et al. 1974; Hood et al. 1982). More recently, EM induction was used to discover liquid water oceans in the icy Galilean satellites of Jupiter (Khurana et al. 1998; Kivelson et al. 1999; Kivelson et al. 2002).

Apollo and Lunar Prospector (LP) data have constrained the radius of a highly conducting lunar core to <400 km (Hood et al. 1999) and determined the deep mantle conductivity (Hood et al. 1982; Hood and Sonnett 1982) and its relation to the geothermal gradient and thermal evolution of the Moon. However, the transfer function is not very well constrained at depths less than 500 km from the surface or radial distances less than 500 km from the center, because at high frequencies the planar approximation breaks down and at low frequencies there are uncertainties in distinguishing the induction signal due to instrument offsets or noise. For example, Explorer 35 data, used to determine the driver in the Apollo era, had significant offset fluctuations, and the lack of simultaneous plasma measurements prohibited identification and removal of ambient space currents. Lunar Prospector studies did not have a nearby monitor of the driver signal.

ARTEMIS will measure the external, driving magnetic field with one probe and the response of the lunar interior to that field with the other probe when it is near periselene. Thus, in a manner analogous to planetary flybys (Khurana et al. 1998; Kivelson et al. 2002), ARTEMIS will determine the response of the conductive core to external field changes. Although the dipole response will be small (0.2–0.8 nT for a driver of dB = 5–20 nT), demonstrated offset stability (<0.1 nT/12 h), noise (<5 pT) and digitization (3 pT) on ARTEMIS/FGM (Auster et al. 2008) enable accurate measurements of the effect. Moreover, the presence of a nearby probe to measure the ambient field including its small variations enables, through subtraction from the total—induced plus external—field measured at periselene, extremely accurate determination of the induced response. Performing such measurements dozens of times over the course of the mission, a database of response as function of position and time relative to the driver impulse, will be assembled, and the core size and conductivity estimated. Differencing the highly sensitive magnetometer signals on the two spacecraft under various external driver frequencies is an ideal way to sound the interior conductivity of the Moon as function of frequency. For the first time the technique will be applied using nearby probes carrying identical sensors with very stable offsets that can be cross-calibrated just hours prior to each pass and can benefit from on-board plasma measurements to remove local space currents.

The ARTEMIS periselene altitude will be less than 100 km (exact altitude depends on results of orbit stability analysis optimizing for planetary goals). This altitude is ideal for making induction measurements from orbit, because with the exception of known, localized magnetic anomalies, all variances from the input signal can be attributed to induction effects. The technique can be applied both in the solar wind at the nightside and in the tail/magnetosheath/lobes on either side of the terminator.

Crustal magnetism. Crustal magnetism preserves ancient records of planetary and surface evolution. At Earth, study of crustal fields revealed polarity reversals of the core dynamo and established a chronology that ultimately confirmed the plate tectonics hypothesis. The origin of lunar magnetism is less clear because of the absence of a present day dynamo. The two strongest anomalies on the near side, Reiner Gamma and Descartes, and the strongest one on the far side, Crisium antipode, have surface fields that likely exceed 1000 nT. Mini-magnetospheric interaction should result in solar wind density enhancements at the front and at the edges of the anomaly, as recently observed by the SELENE/Kaguya ion spectrometer (Saito et al. 2010). These anomalies also provide typical examples of the general correlation between crustal magnetic field regions and high albedo "swirl" features (Richmond et al. 2003; Nicholas 2007). ARTEMIS will measure lunar fields from 100 km or less, depending on the periapsis and longitudes that will be attained, at a 10° inclination or greater (goal $\sim$ 20°), depending on the communications link budget and fuel margin available. It will study the interaction of near-equatorial magnetic anomalies with the solar wind and the magnetotail. These anomalies deflect and shock the solar wind plasma and cause electron heating and wave turbulence (Halekas et al. 2008b, 2008c). Even from 100 km altitude and inclination below 10°, the comprehensive instrumentation on ARTEMIS will measure the magnetic properties of Reiner Gamma and the interaction of this mini-magnetosphere with the solar wind and the Earth's magnetotail.

Thus while ARTEMIS cannot improve upon the geographic coverage of the crustal fields attained by LP and Apollo, the availability of comprehensive *in-situ* instrumentation will greatly expand upon the knowledge gained from prior studies of interactions of such anomalies with the solar wind. For example, ions (including reflected ions) will be measured, the waves from the ion-ion beam instabilities will be sensed in both electric and magnetic fields, and the spacecraft potential from the electric field instrument will be helpful in accurately determining the plasma moments. The high time resolution wave captures will be particularly important in the analysis of plasma waves and in further characterization of the mini-magnetosphere interaction with the solar wind.

Synergies with other missions. Because it overlaps with LRO's extended investigation in 2011 and 2012, the ARTEMIS mission is in a unique position to support LRO's prime and extended mission science objectives. LRO will study the lunar atmosphere and its variability with the LAMP instrument, and particle acceleration mechanisms and their radiation effects on tissue with the CRaTER instrument. ARTEMIS can support LAMP observations of the exosphere by providing accurate measurements of solar wind and magnetotail drivers. Observations during the overlap period between LADEE and ARTEMIS can be used as calibration points to relate the statistical studies that will be done independently by the two missions. CRaTER's objective to study Galactic Cosmic Ray (GCR) and Solar Energetic Particle (SEP) populations will be facilitated by the presence of ARTEMIS as a nearby solar wind monitor.

By measuring both upstream solar wind and local plasma conditions near the Moon, ARTEMIS is also in a unique position to support the Lunar Atmosphere and Dust Environment Explorer (LADEE) mission, slated for a mid-2012 launch. LADEE carries instrumentation to study the dynamics of the lunar exosphere and dust environment, much of which will be tied directly to the ambient plasma conditions at the Moon and in the solar wind. Since LADEE lacks in-situ plasma instrumentation, the presence of ARTEMIS will enable a more direct linkage between specific ambient plasma processes and the resultant exospheric variability measured by LADEE. Moreover, ARTEMIS measurements of the surface potential in tandem with LADEE could revolutionize our understanding of charging processes related to lofted dust that would have gone unnoticed with LADEE measurements alone.

Finally, a major element of NASA's lunar flight projects is the International Lunar Network (ILN), comprised of small geophysical nodes on the lunar surface. These nodes are expected to be deployed in the next decade by NASA and international space agencies. One of the goals of the ILN is to perform lunar EM sounding from the surface with both electric and magnetic sensors. ARTEMIS in orbit will provide continuous magnetometer measurements of the driver signal to meet the needs of the measurement floor of the ILN network's EM sounding goal.

4 Mission Design

The ARTEMIS mission concept originated in 2005, well after the THEMIS Critical Design Review, when it was realized that optimizing the prime THEMIS mission orbit design would result in 8 h-long shadows for P1 and P2 (the outermost THEMIS probes), well beyond their thermal design limits. Since these would occur about six months after the end of the prime mission, alternate plans had to be devised early. Approaches for a mission extended phase were sought in collaboration with NASA/JPL in 2005, when it was realized that by increasing apogee and taking advantage of lunar perturbations, shadows could be avoided with minimal resources, about 100 $\mathrm{m\,s^{-1}}$, although with a complex operations scenario involving station-keeping in translunar orbits. After a re-design of the THEMIS launch vehicle target injection in 2006, which re-optimized the fuel margin on the five THEMIS probes, it was realized that sufficient margins would be available for P1 and P2 at end-of-mission to accommodate a lunar orbit insertion for P1 and P2, assuming a low-thrust injection at the Moon. The lunar orbits had to be highly eccentric, with periods from a few hours to 1.5 days, because there was insufficient fuel margin to accommodate a low-altitude circular orbit. In addition, such orbits were preferred because long periods resulted in infrequent shadows and battery cycling, consistent with the thermal design and verification program of the THEMIS probes. Additionally, the low thrust capability of the probes required that the Lunar Orbit Insertion (LOI) maneuver be split in multiple burns, of which the first was the most critical. Thus, residence in the Lagrange points to properly evaluate and adjust the LOI conditions was deemed necessary in order to reduce operations risk. Since the probes have axial thrusters thrusting only along (but not opposite to) the spin axis direction, and the probes have spin axis approximately along the ecliptic south, it was realized that lunar polar orbits would be less advantageous, as they would result in limited orbit control capability. Finally, lunar orbits had to avoid Earth and Moon shadows longer than 4 h, a revised requirement (relative to a 3 h limit at launch) stemming from the operation team's most recent assessment of the thermal design, based on analysis of in-flight performance data.

In 2007 internal studies at JPL resulted in an initial ARTEMIS trajectory subject to the above constraints that was of sufficient fidelity to be further optimized in collaboration with the science team. In 2008, NASA/HQ requested that the team consider ARTEMIS as part of its 2008 Senior Review process (rather than as a separate proposal) and recommended use of the Deep Space Network (DSN) for data relay, which enabled consideration of the 34 m antennas at a nominal contact frequency of 3.5 h/day. In a series of science working team meetings, the following science considerations were taken in the mission design: (1) The Lissajous orbits were deemed extremely useful scientifically, because they provide information on tail and solar wind spatial scales never measured before. The science team further requested that the Lissajous orbits be performed as two steps, at least 3-months long each: the LL1, 2 and LL1 step (explained in Sect. 1, in particular with Figs. 2 and 3 and Table 1) in order to maximize residence in the large inter-spacecraft separation regime. This was possible by inserting P1 at LL2 and P2 at LL1 for 3 months, followed by bringing P1 at LL1 to have both probes on the near-side of the Moon. (2) The differential precession of the line of apsides of the probes after lunar orbit insertion would be very small if the probes had similar lunar orbital elements. This could place restrictions on the insertion times and strategy and limit observation orientations. To avoid such restrictions and to maximize the inter-probe separation vectors, the science team requested that one of the probes be inserted into a retrograde orbit and the other into a pro-grade orbit to speed up the differential precession of the lines of apsides. P1 was selected for retrograde insertion, to help its fuel margin, since retrograde orbits require less orbit insertion velocity, but this choice may be reconsidered in 2010.

Table 3 Fuel margin available to execute ARTEMIS for heliophysics science goals; additional fuel may be required for modifications and maintenance corresponding to planetary objectives. DSM = Deep Space Maneuver. LOI = Lunar Orbit Injection. TCMs = Trajectory Correction Maneuvers

ARTEMIS Mission (P1, P2): ΔV overview				
Phase	Interval	Maneuver	dV P1	dV P2
TLI	Oct.09–	Orbit raise, Lunar fly-by	100.7	185.6
	Oct.10	Declination, Gravity Losses	20	28
LL1,2	Oct.10–Jan.11	DSM	0.9	17.3
LLI	Jan. 11	Maintenance	15	12
LO	Apr.11–	Lunar transfer initiation, LOI	86.6	108.7
	Sep.12	Decl., Gravity, Steering	7	12
all	all	TCMs	13	10
Total required for this ARTEMIS probe			243	374
Total available at end of prime mission			300	450
ΔV available ARTEMIS margin [m/s]			57	76
ΔV available for ARTEMIS margin (%)			23%	20%

(3) Equatorial, highly eccentric lunar orbits of ~26 h period (100 km × 18,000 km altitude) were deemed most useful scientifically for the lunar orbit phase, as they enable separations of up to 5 R_E (18 R_L) at all orientations over the course of 2 years. Higher aposelenes would have caused increased Earth perturbations that would have resulted in early orbit insertion. Additionally, higher aposelenes would have also produced longer lunar shadows (beyond the 4 h limit requirement).

These science desires and mission operational constraints were worked into the final orbit scenario described in Sect. 1. In particular, the near-equatorial orbits, of period ~1 day are easy to achieve with the side-thrusting capability (which allows thrusting at any vector orientation along the spin plane). Sufficient margin is available at the end of the nominal ARTEMIS mission design, as shown in Table 3. Since lunar and other perturbations also necessitate correction maneuvers that may be out of the spin plane, care must be taken to ensure that the return-to-nominal plan of the ARTEMIS operations team is achieved by the axial thrusters for reasonable (3 sigma) deviations of the trajectory from nominal. This is done by the ARTEMIS navigation team at GSFC, which analyzes and biases the nominal orbit such that achievable correction maneuvers can be inserted at specific points into the mission, if deemed necessary based on the actual maneuver execution and thruster performance in orbit. The resultant ARTEMIS lunar orbits are very similar, operationally, to the ones for the THEMIS mission at Earth, in terms of a thermal environment, power cycling, communications plan, and data collection strategy. These are dictated primarily by the properties of the equatorial, highly eccentric, nearly day-long period orbits, resulting in shadows that are below the four-hour flight-demonstrated extended survival limits of the probes during their Earth-orbit history. In addition to lunar shadows, care must be taken to predict and avoid ARTEMIS Earth shadows through mean anomaly phasing; this costs very little fuel if achieved far in advance, else it may result in shadows that can exceed the design limit. This combination of science and technical trades has resulted in a robust, low-risk mission design solution for ARTEMIS, and is expected to provide an unprecedented view of the lunar space environment in a very cost-effective way.

In late 2008 it was realized that significant planetary goals can also be achieved from ARTEMIS, with only small modifications to the mission design. These are a periselene altitude reduction and an inclination increase. With regards to the periselene altitude, the P1, P2 orbits are expected to be further optimized to "graze" the surface in the < 100 km domain once a month. By expending maintenance fuel on the order of a few m s^{-1} (see Table 3 for a perspective with regards to fuel margins), it is possible to maintain a stable orbit at low periselene. Considerations will be given to the fuel margins prior to orbit insertion and the operations complexity from periselene maintenance in the remainder of the mission. With regard to inclination adjustments, an inclined orbit results in additional opportunities of conjunctions with crustal anomalies near periapsis. But an inclined orbit increases the gravity gradient torque on the spin axis away from its optimal orientation of 3–13° (8° nominal). This affects communications as there are significant signal losses below 15° from the spin plane. Additional considerations include thermal effects, boom shadow effects on instrument performance, and station-keeping fuel. An inclination between 10–20° is expected to be achievable. The exact value will be determined closer to insertion time. These planetary science optimizations will be revisited in the summer of 2010, after translunar injection, and the results will be folded into the ARTEMIS mission design, assuming sufficient resources are available, in early 2011.

5 ARTEMIS Mission Operations Plans

The ARTEMIS mission is, by design, a natural evolution from THEMIS operations at Earth to operations in the lunar environment, in terms of spacecraft commanding and conditioning, instrument modes, instrument operations, and data relay/processing strategy. The ARTEMIS Mission Operations Systems is comprised of Mission Operations and Science Operations following the practices of THEMIS: ARTEMIS is operated by the Mission Operations Center (MOC) at the Space Sciences Laboratory, University of California, Berkeley (Bester et al. 2008). In addition, the JPL mission design team has developed and delivered the mission trajectory to the MOC for implementation; the GSFC flight dynamics team has developed for ARTEMIS Sun-Sensor—only attitude determination solutions; and the GSFC navigation group supports ARTEMIS in performing navigation error analysis, inserting orbit biases, and determining trajectory correction maneuvers that need to be inserted to compensate for a return-to-nominal mission design plan. The mission operations center performs: mission planning functions in accordance with science (instrument operation modes) requests; flight dynamics; orbit and attitude determination; maneuver planning; commanding and state-of-health monitoring of the five probes; recovery of science and engineering data; data trending and anomaly resolution. Science operations comprise the generation of instrument schedules, data processing and archiving, generation and maintenance of data analysis and display software, instrument trending, and science community support.

The main operations differences between ARTEMIS and THEMIS are: (i) the use of the Deep Space Network's (DSN) 34 m antennas for communications, and (ii) the instrument operations that will have to be adjusted to the new environment, reduced data volumes, and new science. Use of the DSN antennas necessitated new operational interfaces (ephemeris, scheduling, telemetry/command/tracking data, and file transfers) and processing/conversion tools. The integration of the DSN antennas into the existing MOC network is seamless, however. For example, range and range-rate data from the DSN are transmitted to the MOC, translated into the same format as the Berkeley Ground Station and the rest of the NASA ground network, and processed with the standard tools (GTDS) for orbit determination. The

Table 4 Typical mode of operations of ARTEMIS probes for Heliophysics and Planetary investigations, per 2 orbits. An orbit is ~26 h. Bursts will be triggered either based on-board triggers, or based on time. FIT = E, B spin fits; MOM = ESA moments; RDFs = reduced particle distributions; FDFs = Full (angular) particle distributions; FBK = filterbank wave spectra; FFFs = Fourier wave spectra

Mode	Duration	Products
Slow Survey	2 orbits	FIT, MOM, RDFs, FBK
Fast Survey	3 h	FDFs, FFFs, Waveforms
PB (2/orbit)	40 min	Full cadence FDFs
WB(2 per PB)	2×6 s	Full cadence waveforms

main effect is that given the DSN link margin and contact duration, approximately $1/5$ of the amount of data that can be recorded on memory can be transmitted to the ground per contact (nominally once per two days, per probe). (The nominal instrument modes have been discussed in Angelopoulos (2008).) Because there is one contact per probe over the period of 2 days, careful planning and selection of the Fast Survey (FS), Particle Burst (PB), and Wave Burst (WB) intervals are required (Table 4). The downlink volumes are consistent with one FS interval of only ~1.5 h per orbit (compared to 12 h per orbit at Earth) and the selection of one to two (max) PB per orbit with one or two (max) embedded WBs within it. The location of FS is time-based and will be one of the following: wake crossing, periselene, two probe wake/boundary alignment, nominal plasma sheet, or boundary layer crossing (magnetotail). Since not all of those can be achieved simultaneously, the operations plan involves mission phases to optimize data collection for specific science objectives at various parts of the mission. Alternative collection plans are also currently being considered.

The second difference is changes to instruments to best suit the proposed studies in the new environment and commanding of the instrument modes to obtain the optimal heliophysics and planetary science. ARTEMIS cares about instrument sensitivity far beyond the requirements of THEMIS, because it operates in a 10 nT typical ambient field and near-background particle fluxes, except for the solar wind beam population. With regard to the FGM and SCM instruments, cleanup methods have already been devised and disseminated (though not very widely used yet). Their operational use and efficacy and any changes in response to community feedback remain tasks for the future.

For the FGM instrument these include corrections to: ground processing of data around range changes, digital-to-analog differential non-linearity effects, offset drifts, and power system currents. Since it is anticipated that the magnetometer will be in a single range (likely range 8, ~3 pT resolution, ~100 nT maximum field, see Auster et al. 2008), and the non-linearity affects the data in high fields only, by far the most important corrections necessary in the lunar phase are the offset drift correction and the power system current noise removal. Routine offset determination using recently developed techniques in the solar wind (Leinweber et al. 2008) and calibration of L2 data to account for power system currents will be done by the science team in the operational phase of ARTEMIS. For the ESA instrument (McFadden et al. 2008a, 2008b), the solar wind mode routinely used on THEMIS is incompatible with the need to measure simultaneously the upstream ions, wake-accelerated or solar wind shock-accelerated ions, or lunar backscattered ions. The current plan is therefore to utilize the electron sensor for determining solar wind velocity, density, and electron temperature and ignore the ion saturation that occurs from the nominal magnetospheric mode operation of the ion sensor. The inter-calibration of ESA electron density and velocity with the other instruments (electric field, ion ESA) is currently under way. An alternative approach is to create a hybrid instrument operational mode with higher angle resolution in the solar wind direction, preserving the full energy resolution achievable in the magnetospheric mode. This approach will enable solar wind ion density and velocity measurements and provide reasonable total ion temperature measurements, as well.

Additional effects from shadows on all instruments are present. In the absence of a sunpulse the spin period is not known internally on the spacecraft, and particle instrument sectoring of distribution functions and on-board spin-fits of the magnetic and electric field data become suspect. As the spin period changes due to thermal effects (wire boom contraction causes probe to spin up), the above sectoring and spin-fit algorithm results in products with an apparent drift in sun-angle. The drift can result in several full rotations per hour, but since the spinup is a combination of several non-linear terms (as it depends on amount of partial sun illumination, wire/probe bus heat capacity, and moment of inertia), it can only be fitted to observed data. This has been done with good success using the FGM data, and routines for correcting the shadow data are now available. The spin-phase information will be included as a correction to the spin-phase files in the post-processing tools for all particles and fields quantities to produce accurately post-processed distributions and properly oriented spin-fits.

Science data processing, calibration, dissemination, and analysis tools development and maintenance follow the successful practices of THEMIS. Level-0 data are uncalibrated raw data, in day-file and raw packet format. Level-1 data are Common Data Format (CDF) files containing uncompressed, time-ordered, overlap-deleted data in raw instrument units (e.g., counts, 16 bit integers). They are efficiently packaged, and can be read by any platform that supports the NASA/GSFC-distributed CDF platform. L1 files are read automatically by IDL-based, freely and widely distributed analysis tools and are calibrated on-the-fly using the latest calibration parameters to take advantage of the most updated calibration done on the instrument without the need for any L1 file reprocessing. L2 files contain calibrated data of a subset of the dataset, representing the most important quantities from each instrument. L2 files are in physical units and also in CDF format and do not require further calibration; they can be read by any software that is able to access CDF files, such as Fortran, C, Matlab, and IDL. They are simple enough to be easily interpreted using standard documentation. Automated processing performs both standard calibration (using the latest parameters and orbit-predicts) and L2 file production within hours of data receipt at the MOC. Standard overview plots are also automatically produced to facilitate data quality evaluation and quick event selection, especially in conjunction with other missions. An example of an overview plot is shown in Fig. 7. Events such as calibration file changes or definitive orbit/attitude file updates trigger reprocessing of both L2 files and on-line plots.

Files are disseminated via project site and mirror site web pages by NASA/GSFC's Space Physics Data Facility and by Virtual Observatories. Plans for inclusion of ARTEMIS data into the Planetary Data System (PDS) are currently under way. The most useful and more prevailing means of data dissemination, however, is via the THEMIS and ARTEMIS data analysis software, an IDL-based suite of reading, analysis, and visualization routines offering both command-line and Graphical User Interface capability. This code, used by the science team, is freely available to the science community. Data ingestion is performed "on-the-fly", along with calibration, if L1 quantities are being introduced. The code interrogates the user's preferred http or ftp sites (the default is project site) and downloads only the data that has been reprocessed recently, based on the file creation date at the user's machine relative to the remote site. Batch downloads or bundled downloads are also possible but not required or needed. Once the data resides on a user's machine, analysis is possible off-line. Dozens of crib sheets, i.e., text files containing IDL code that demonstrates usage of certain routines, are available with the data distribution. Special-purpose crib sheets are also

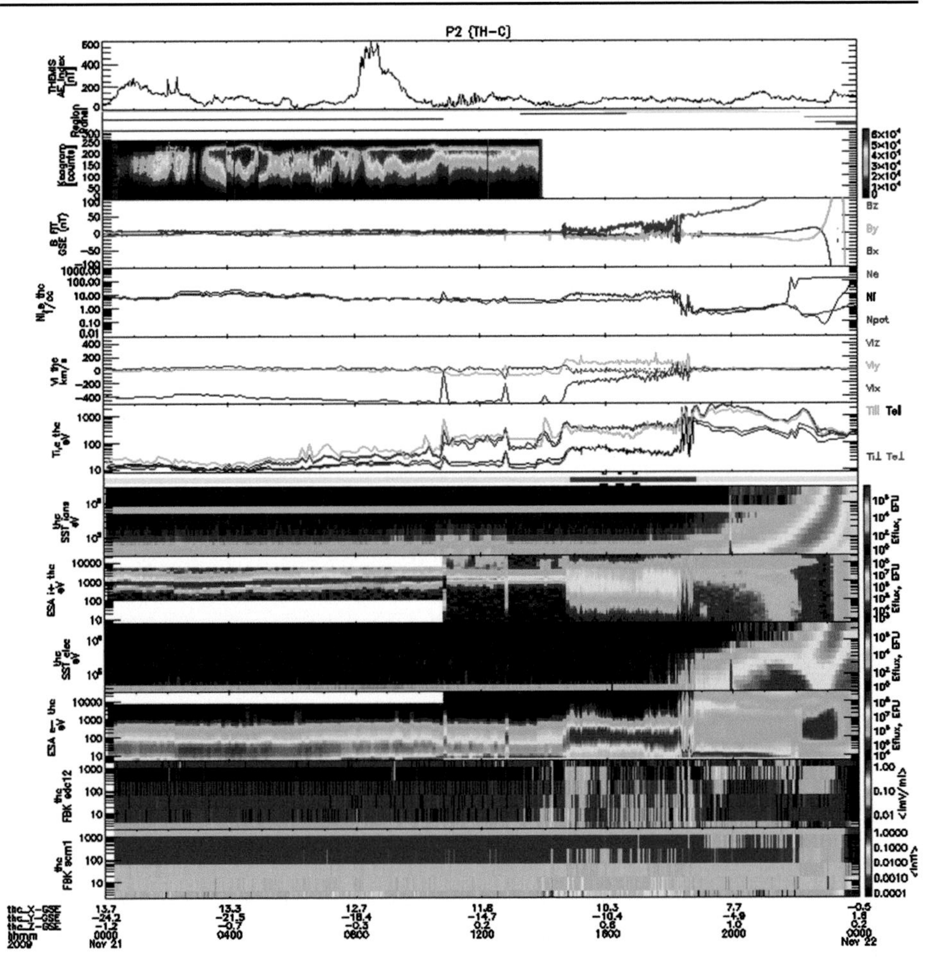

Fig. 7 Overview plot (routinely produced) from all instruments on ARTEMIS probe P2 during a perigee passes a few months prior to its translunar injection. Horizontal bar indicates mode (SS = yellow, FS = red, PB/WB = black bottom/top)

exchanged between users, enabling rapid, efficient communication and exchange of tools, experience, and ideas. The analysis code distributed also provides a Graphical User Interface (GUI) that allows users completely unfamiliar with command line IDL coding to have both quick access to the data and a fast introduction into the ARTEMIS analysis system (Fig. 8). The GUI is also accessible by IDL's product: "Virtual Machine", which is free of charge and also contains data manipulation capability by virtue of a "mini-language" operating on data structures or on arrays. Furthermore, the GUI allows easy plot manipulation capability (panning, zooming, line colors, symbol fonts, plotting), permitting publication-quality plots. The IDL calibration and analysis code is disseminated to the community via the THEMIS/ARTEMIS web sites; tutorials are routinely conducted at various institutions and during major international meetings.

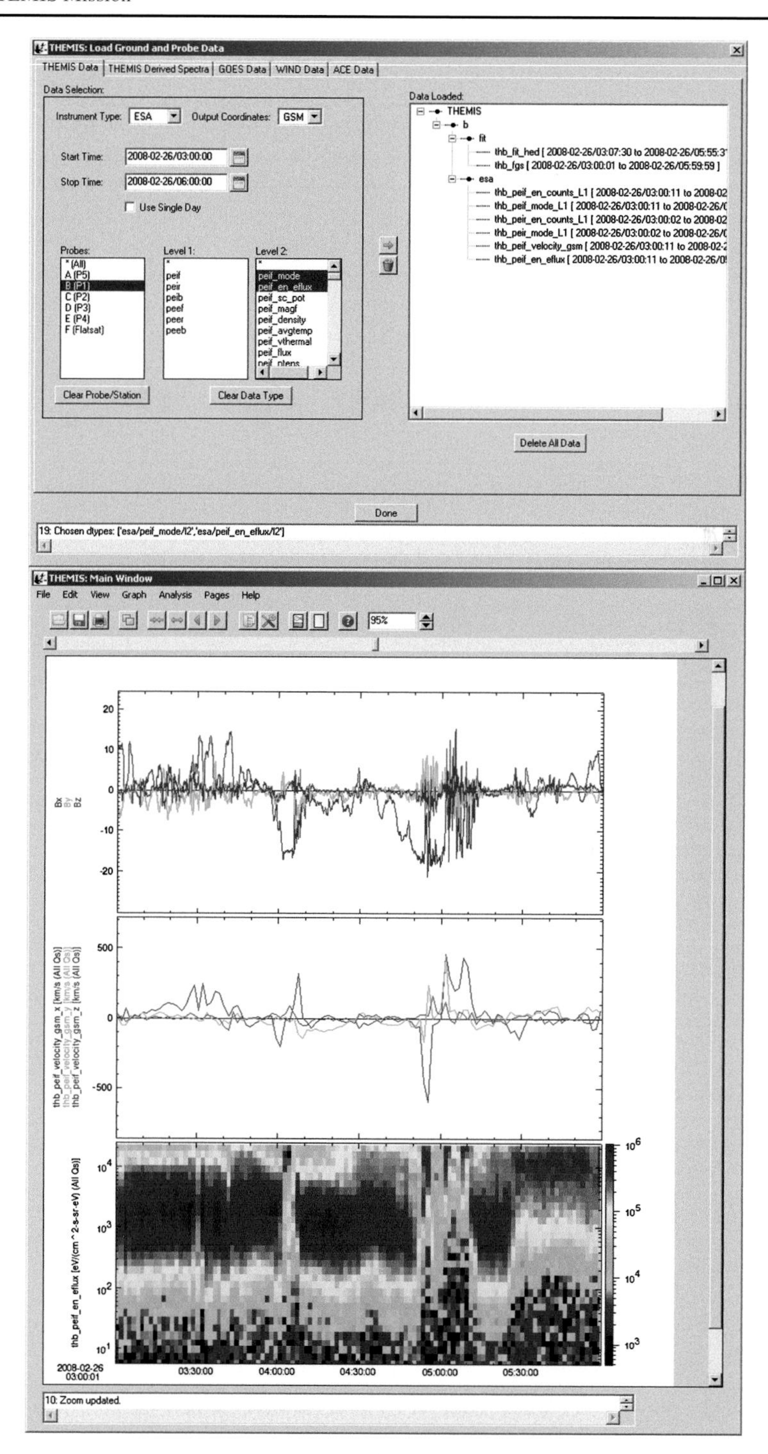

Fig. 8 THEMIS/ARTEMIS Graphical User Interface showing Data Introduction panel and Plotting tool. The GUI benefits from the underlying command line software but provides tools for both plot manipulation and for easy manipulation of data quantities, seamlessly easing novice users into the data system

6 Summary

ARTEMIS, the first two micro-satellite mission to the Moon, is a cost-effective interdisciplinary mission addressing fundamental questions in NASA's Heliophysics and Planetary Disciplines. ARTEMIS will study space physics processes in the solar wind, magnetotail, and the lunar wake, and can also address important questions on the exospheric, surface, and subsurface lunar environment. With stable orbits, ARTEMIS can monitor both the solar wind and the lunar environment during the next solar cycle; it is therefore aligned with NAC/NRC's recommendations to characterize the lunar environment and predict space weather impacts on robotic and human productivity. Finally, being the first mission to use prolonged residence in lunar libration orbits, which are important for communications and as staging grounds for lunar landings, ARTEMIS represents a pathfinder for future lunar exploration missions.

Acknowledgements ARTEMIS was made possible thanks to the excellence of the implementing teams and the tenacity of the many scientists "over the Moon" at the expectation of its scientific returns. The project is blossoming into both an exploratory mission of and from the lunar environment, and a dependable, long term node of the Heliophysics Observatory. Dr. D. Sibeck was instrumental in the early mission design and a constant source of encouragement; he organized the science deliberations that produced the first distilled, pragmatic, yet exciting scientific outline of the mission plan. Heartfelt thanks to Drs. D. Brain, G. Delory, J. Eastwood, W. Farrell, R. Grimm, J. Halekas, H. Hasegawa, K. Khurana, D. Krauss-Varban, R. Lillis, M. Oieroset, T. Phan, J. Raeder, C. Russell, P. Travnicek, J. Weygand, D. Schriver, and J. Slavin, who helped define a focused and highly optimized scientific investigation, and to Drs. H. Spence, N. Fox, P. Brandt, M. Collier, and D. Rowland for their careful review and constructive criticism of the concept while still in its infancy; the discipline owes you for your gracious contribution of your time. The JPL mission design team was creative and patient with the scientists' requests; in particular, S. Broschard, M. K. Chung, S. Hatch, J. Ma, T. Sweetser, and G. Whiffen did a fantastic job that has stood the test of reviews and time—they are the first of the team to claim "victory". Many thanks to the GSFC navigation team, D. Folta, M. Woodard, and D. Woodfork, for validating the "return-to-nominal" concept is achievable with less-than-3D thrust capability, for building up the navigation error and maneuver correction plan, and for their continued support with trajectory optimizations and maneuver execution in the translunar and Lissajous phases. Thanks also go to J. Hashmall, D. Felikson, and J. Sedlak for developing the Sun-only attitude solutions that enabled ARTEMIS to find its way without a compass. I owe my deepest gratitude to the ARTEMIS Mission Operations Team at UC Berkeley for having operated the two ARTEMIS probes flawlessly, conserving fuel margin despite the 2nd year THEMIS science re-optimization. The team, led by Dr. M. Bester, is comprised of M. Lewis, S. Frey, D. Cosgrove, B. Roberts, J. McDonald, D. Pease, J. Thorsness, J. Marchese, B. Owens, S. Gandhi, M. Eckert, R. Dumlao, G. Lemieux, G. Picard, S. Johnson, and T. Clemons. They are the quiet heroes behind the many discoveries sure to come. After turning every last drop of hydrazine available on THEMIS margin into new capabilities for ARTEMIS, they rolled up their sleeves and got to work on a challenge even more ambitious than simultaneous maneuvering of five spacecraft with a combined 100 maneuvers per year (who would have imagined there could have been one?). Their careful post-maneuver data analyses, which lead to a thruster performance prediction model, have substantially reduced operations risk, providing us a safer ride to the final ARTEMIS destination. A great many thanks go to the spacecraft provider, ATK Space (formerly SWALES Aerospace Inc.), for a well-behaved set of probes; and to an experienced and committed instrument team for building, calibrating, and maintaining five impeccable sets of instruments—they have made operations easier and enabled the full panoply of THEMIS to be deployed in a new journey of exploration and discovery. ARTEMIS was made possible by NASA under contract NAS5-02099.

References

M.H. Acuña et al., The Global Geospace Science program and its investigations. Space Sci. Rev. **71**, 5 (1995)
V. Angelopoulos, The THEMIS Mission. Space Sci. Rev. (2008). doi:10.1007/s11214-008-9336-1
H.U. Auster et al., The THEMIS fluxgate magnetometer. Space Sci. Rev. (2008). doi:10.1007/s11214-008-9365-9
M. Bester et al., THEMIS operations. Space Sci. Rev. (2008). doi:10.1007/s11214-008-9456-7
A.B. Binder, Lunar Prospector, overview. Science **281**, 1475 (1998)

J.W. Bonnell et al., The Electric Field Instrument (EFI) for THEMIS. Space Sci. Rev. (2008). doi:10.1007/s11214-008-9469-2
S.B. Broschart et al., Preliminary trajectory design for the ARTEMIS lunar mission. AAS 09-382, 2009
C.M. Cully et al., The THEMIS digital fields board. Space Sci. Rev. (2008). doi:10.1007/s11214-008-9417-1
P. Dyal et al., Magnetism and the interior of the Moon. Rev. Geophys. Space Phys. **12**, 568 (1974)
J.S. Halekas et al., Extreme lunar surface charging during solar energetic particle events. Geophys. Res. Lett. **34**, L02111 (2007). doi:10.1029/2006GL028517
J.S. Halekas et al., Lunar Prospector observations of the electrostatic potential of the lunar surface and its response to incident currents. J. Geophys. Res. **113**, A09102 (2008a). doi:10.1029/2008JA013194
J.S. Halekas et al., Density cavity observed over a strong lunar crustal magnetic anomaly in the solar wind: A mini-magnetosphere? Planet. Space Sci. **56**, 941 (2008b). doi:10.1016/j.pss.208.01.008
J.S. Halekas et al., Solar wind interaction with lunar crustal magnetic anomalies. Adv. Space. Res. **41**, 1319 (2008c). doi:10.1016/j.asr.2007.04.003
J.S. Halekas et al., Lunar surface charging during solar energetic particle events: Measurement and prediction. Planet. Space Sci. **57**, 78 (2009)
R.E. Hartle, R. Killen, Measuring pickup ions to characterize the surfaces and exospheres of planetary bodies: Applications to the Moon. Geophys. Res. Lett. **33**, L05201 (2006). doi:10.1029/2005GL024520
P. Harvey, E. Taylor, R. Sterling, M. Cully, The THEMIS constellation. Space Sci. Rev. (2008). doi: 10.1007/s11214-008-9416-2
L.L. Hood, C.P. Sonnett, Limits on the lunar temperature profile. Geophys. Res. Lett. **9**(1), 37 (1982)
L.L. Hood et al., The deep lunar electrical conductivity profile—structural and thermal inferences. J. Geophys. Res. **87**, 5311 (1982)
L.L. Hood et al., Initial measurements of the lunar induced magnetic dipole moment using lunar prospector magnetometer data. Geophys. Res. Lett. **26**(15), 2327 (1999)
G.S. Hubbard et al., The Lunar Prospector discovery mission: mission and measurement description. IEEE Trans. Nucl. Sci. **3**, 880 (1998)
K.K. Khurana et al., Induced magnetic fields as evidence for subsurface oceans in Europa and Callisto. Nature **395**, 777–780 (1998)
M.G. Kivelson et al., Europa and Callisto: Induced or intrinsic fields in a periodically varying plasma environment. J. Geophys. Res. **104**, 4609 (1999)
M.G. Kivelson et al., The permanent and inductive magnetic moments of Ganymede. Icarus **157**, 507 (2002). doi:10.1006/icar.2002.6834
O. Le Contel et al., First results of the THEMIS searchcoil magnetometers. Space Sci. Rev. (2008). doi:10.1007/s11214-008-9371-y
H.K. Leinweber et al., An advanced approach to finding magnetometer zero levels in the interplanetary magnetic field. Meas. Sci. Technol. **19**(5), 055104 (2008)
J.P. McFadden et al., The THEMIS ESA plasma instrument and in-flight calibration. Space Sci. Rev. (2008a). doi:10.1107/s11214-008-9440-2
J.P. McFadden et al., THEMIS ESA first science results and performance issues. Space Sci. Rev. (2008b). doi:10.1007/s11214-008-9433-1
J.B. Nicholas, Age spot or youthful marking: Origin of Reiner Gamma. Geophys. Res. Lett. **34**, L02205 (2007). doi:10.1029/2006GL027794
A. Nishida, The GEOTAIL mission. Geophys. Res. Lett. **21**, 2871 (1994)
M.N. Nishino et al., Solar-wind proton access deep into the near-Moon wake. Geophys. Res. Lett. **36** (2009). doi:10.1029/2009GL039444
W.D. Parkinson, *Introduction to Geomagnetism* (Scottish Academic, Edinburgh, 1983)
T.D. Phan et al., A magnetic reconnection X-line extending more than 390 Earth radii in the solar wind. Nature **439**, 175 (2006)
N.C. Richmond et al., Correlation of a strong lunar magnetic anomaly with a high-albedo region of the Descartes mountains. Geophys. Res. Lett. **30**(7), 1395 (2003). doi:10.1029/2003GL016938
A. Roux et al., The search coil magnetometer for THEMIS. Space Sci. Rev. (2008). doi:10.1007/s11214-008-9455-8
C.T. Russell et al., Magnetic evidence for a lunar core. Proc. LSC **12**, 831 (1974)
Y. Saito et al., In-flight performance and initial results of Plasma energy Angle and Composition Experiment (PACE). Space Sci. Rev. (2010). doi:10.1007/s11214-010-9647-x
D.G. Sibeck, V. Angelopoulos, THEMIS science objectives and mission phases. Space Sci. Rev. (2008). doi:10.1007/s11214-008-9393-5
Sweetser et al., ARTEMIS mission design. Space Sci. Rev. (2010, this issue)
B.T. Tsurutani, T.T. von Rosenvinge, ISEE-3 distant geotail results. Geophys. Res. Lett. **11**, 1027 (1984)

DOI 10.1007/978-1-4614-9554-3_3
Reprinted from *Space Science Reviews* Journal, DOI 10.1007/s11214-011-9777-9

ARTEMIS Science Objectives

D.G. Sibeck · V. Angelopoulos · D.A. Brain · G.T. Delory · J.P. Eastwood · W.M. Farrell · R.E. Grimm · J.S. Halekas · H. Hasegawa · P. Hellinger · K.K. Khurana · R.J. Lillis · M. Øieroset · T.-D. Phan · J. Raeder · C.T. Russell · D. Schriver · J.A. Slavin · P.M. Travnicek · J.M. Weygand

Received: 25 July 2010 / Accepted: 8 April 2011 / Published online: 17 May 2011

Abstract NASA's two spacecraft ARTEMIS mission will address both heliospheric and planetary research questions, first while in orbit about the Earth with the Moon and subsequently while in orbit about the Moon. Heliospheric topics include the structure of the Earth's magnetotail; reconnection, particle acceleration, and turbulence in the Earth's magnetosphere, at the bow shock, and in the solar wind; and the formation and structure of the

D.G. Sibeck (✉)
Code 674, GSFC/NASA, Greenbelt, MD 20771, USA
e-mail: david.g.sibeck@nasa.gov

V. Angelopoulos
IGPP, UCLA, Los Angeles, CA 90095, USA

D.A. Brain · G.T. Delory · J.S. Halekas · R.J. Lillis · M. Øieroset · T.-D. Phan
UCB, Berkeley, CA, USA

J.P. Eastwood
Imperial College, London, UK

W.M. Farrell · J.A. Slavin
NASA/GSFC, Greenbelt, MD, USA

R.E. Grimm
SWRI, Boulder, CO, USA

H. Hasegawa
ISAS, Sagamihara, Japan

P. Hellinger · P.M. Travnicek
Astronomical Institute, Prague, Czech Republic

K.K. Khurana · C.T. Russell · D. Schriver · P.M. Travnicek · J.M. Weygand
UCLA, Los Angeles, CA, USA

J. Raeder
UNH, Durham, NH, USA

lunar wake. Planetary topics include the lunar exosphere and its relationship to the composition of the lunar surface, the effects of electric fields on dust in the exosphere, internal structure of the Moon, and the lunar crustal magnetic field. This paper describes the expected contributions of ARTEMIS to these baseline scientific objectives.

Keywords ARTEMIS · Moon · Reconnection · Particle acceleration · Turbulence · Wake · Lunar surface · Lunar core · Dust · Electric fields · Crustal anomalies

1 Introduction

ARTEMIS (Acceleration, Reconnection, Turbulence, and Electrodynamics of the Moon's Interaction with the Sun) is NASA's first dual spacecraft mission to the Moon. By mission design and both efficient navigation and flight operations during the primary phase of the THEMIS mission from February 2007 to September 2009 (Angelopoulos 2008), the outermost two THEMIS spacecraft P1 and P2 found themselves with fuel reserves sufficient to undertake a series of over forty maneuvers, including multiple lunar approaches and flybys, that enabled them to reach and parallel the Moon during trans-lunar injection phases from October 2009 through September 2010. From October 2010 to January 2011, the two spacecraft were in Lissajous orbits at the Lagrange points of the Earth-Moon system that straddled the Moon and were separated by distances on the order of $\sim$20 Earth radii (R_E). From January to July 2011, their Lissajous orbits were separated by 5 to 20 R_E on the same side of the Moon. In July 2011 they entered lunar orbits (one prograde, the other retrograde) with separation distances ranging from 500 km to 20 lunar radii (R_L). Note that 1 R_E is 6371 km, while 1 R_L is 1737 km.

Observations from lunar orbit over a such wide range of interspacecraft separations are ideal for addressing a number of longstanding heliophysics objectives, including defining the characteristics of reconnection and turbulence in the solar wind and distant magnetotail, the nature of particle acceleration at the Earth's bow shock and interplanetary discontinuities, and the electrodynamics of the solar wind's interaction with the Moon. However, by careful optimization of mission parameters, in particular periselene and orbital inclination, the fully-instrumented and fully-functional ARTEMIS spacecraft can also be employed to address a set of planetary research objectives including the composition and magnetization of the lunar surface, the effects of electric fields on dust in the vicinity of the Moon, and the structure of the lunar interior. With orbits that will remain stable for over a decade, ARTEMIS will provide observations of the lunar environment over solar cycle time scales, offer measurements that overlap and enhance those of NASA's Lunar Reconnaissance Orbiter (LRO) and Lunar Atmosphere and Dust Environment Explorer (LADEE) missions, and provide crucial solar wind observations for correlative studies with other NASA missions to the Earth's magnetosphere and ionosphere/thermosphere.

Angelopoulos (2011, this issue) provides a mission overview for ARTEMIS that summarizes the overall scientific objectives, orbits, instrumentation, mission and science operations that are expanded upon in individual papers in this compendium. This paper describes how the above mission elements will enable the ARTEMIS team to reach scientific closure. In terms of organization, Sect. 2 describes ARTEMIS planetary objectives while Sect. 3 describes the ARTEMIS heliophysics objectives. Section 4 presents conclusions.

2 ARTEMIS Planetary Science Objectives

The Moon holds critical information regarding the origin of the solar system. Because it is devoid of plate tectonics, volcanism and surface-altering atmospheric processes, the Moon's surface has recorded the 4.5 billion years of solar system history more purely than any other planetary body. Understanding the lunar surface and the stratification of the lunar interior provides a window into the early history of the Earth-Moon system and can shed considerable light on the evolution of terrestrial planets such as Mars and Venus. The surface is constantly sputtering ions and dust, which then enter and circulate within the lunar exosphere before escaping into the solar wind. Ions 'picked-up' by the solar wind gain similar velocities but mass-dependent energies as they follow mass-dependent trajectories. ARTEMIS will use its energy-angle spectroscopic capability and its electric and magnetic field sensors to determine the motion, source, composition, and flux of exospheric ions. The surface electric fields that loft and transport dust particles can be remotely sensed using electron reflectometry or measured directly from altitude. ARTEMIS will use its electromagnetic field measurements to determine the forces acting on dust populations and cause their acceleration and deposition or loss in the lunar environment.

Apollo-era data indicates that the Moon formed via the impact of a Mars-sized object with the early Earth, and later differentiated into primary crust, mantle residuum, and possibly a small iron-rich core. Radial profiles for the Moon's temperature and composition, and their lateral variability today, hold important clues regarding the history of lunar differentiation. Broadband ($\ll$10 mHz to 10 Hz) electromagnetic sounding will improve our knowledge of these state variables for the core, mantle, and crust. The sensitive magnetometers on ARTEMIS will approach the Moon on orbits that bring them, one at a time, to within 100 km of the lunar surface, i.e. to altitudes close enough to detect the core response to varying external drivers but distant enough to minimize perturbations from crustal anomalies. ARTEMIS also provides sensitive horizontal electric field measurements from 1–10 Hz that will further enable magnetotelluric investigations.

Finally, ARTEMIS will use its comprehensive particle and field sensors to study the interaction of the solar wind with lunar crustal magnetic anomalies. Initial Kaguya observations have already provided significant new information on an ion sheath, electron heating and solar wind reflection around magnetic anomalies at 100 km (Saito et al. 2010), but wave properties and solar wind particle flow around the strong field region remain poorly understood. ARTEMIS will study the magnetic anomalies to infer properties of the ancient, seed magnetic field and to determine the accessibility of the solar wind to the surface and the effect it has on lunar surface ageing. Electric field and plasma wave data, together with ion and electron measurements in the vicinity of the mini-magnetosphere and shock-like structures (e.g., Halekas et al. 2006a) that form around the crustal anomaly (the first comprehensive plasma measurements attempted at the Moon) promise exciting new science with possibly significant ramifications for planetary evolution.

2.1 Exospheric Ions and Plasma Pick-up

Forty years after the first Apollo landings, the composition and structure of the lunar exosphere remain poorly understood. A number of studies have used ground-based measurements to investigate the existence, extent, variability and likely sources of easily observed exospheric species such as sodium and potassium (Potter and Morgan 1988; Tyler et al. 1988; Mendillo et al. 1991, 1999; Wilson et al. 2003, 2006). However, the behavior of these species may not be representative of other, more abundant, species (Potter and Morgan 1988; Flynn and Mendillo 1993).

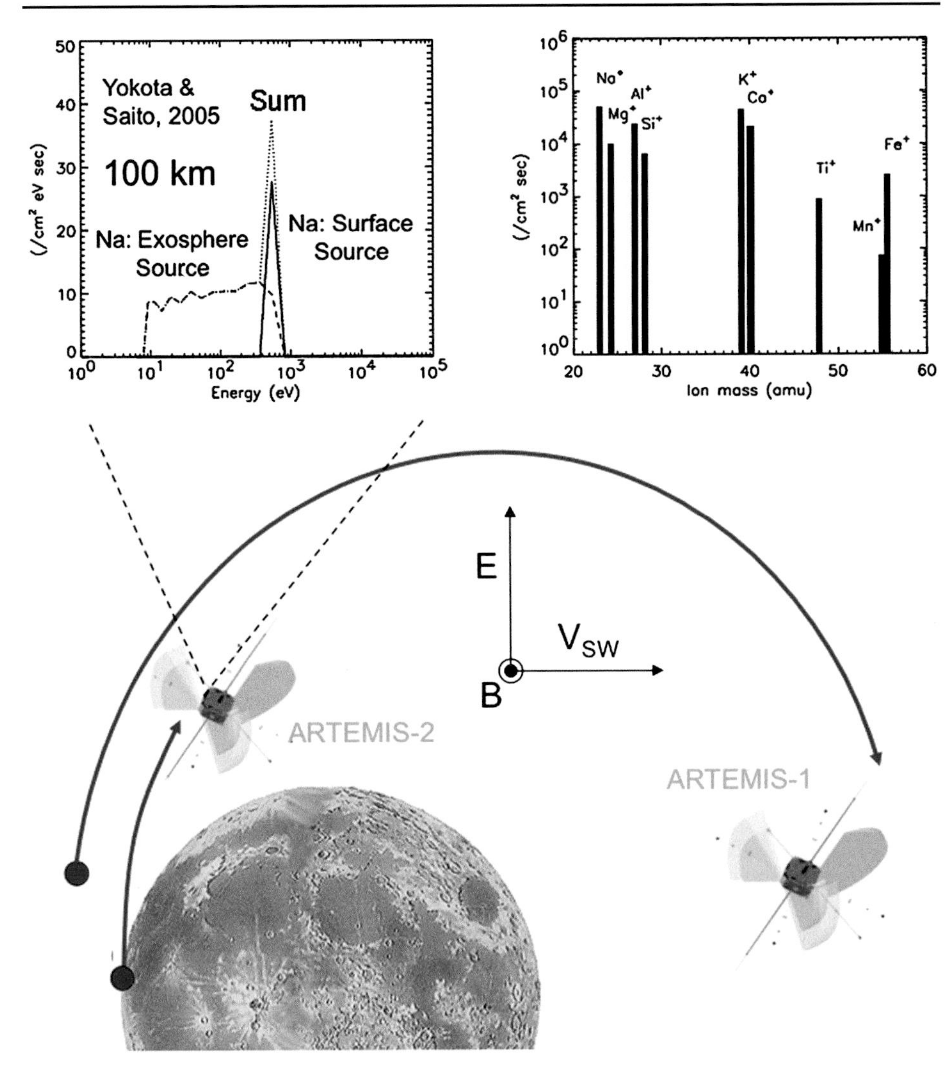

Fig. 1 A schematic showing the trajectories of recently picked-up ions and the expected measurements of the fluxes (*top left*) and composition (*top right*) of the picked-up ions

In situ measurements of pickup ions offer an appealing complementary means of probing surface and exospheric properties and processes at the Moon (e.g. Hartle and Thomas 1974; Cladis et al. 1994; Yokota and Saito 2005; Hartle and Killen 2006; Hartle and Sittler 2007; Saito et al. 2010). Newly created ions, produced by solar wind sputtering, photostimulated desorption, micro-meteoriod impact vaporization or ionization of exospheric gases are generated at relatively low energies (0.01–10 eV), but immediately feel the effect of solar wind magnetic and electric field (Stern 1999; Wurtz et al. 2007). The 'picked-up' ions are then accelerated in cycloidal trajectories like that shown in Fig. 1.

In situ measurements by ion instruments on the space plasma missions AMPTE, WIND, and GEOTAIL confirmed the presence of lunar heavy ions far downstream from the Moon in the solar wind and in the distant magnetotail (Hilchenbach et al. 1991, 1993; Kirsch et al. 1997; Mall et al. 1998), but relied on fortuitous configurations of the solar wind and

IMF and were not able to provide comprehensive information on exospheric composition, source rates, or variability. Further, these studies could not distinguish between lunar ions originating at the surface and those created in the exosphere.

More recent studies have shown that it is possible to separate the signatures of surface and exospheric ions, using measurements made at altitudes smaller than the typical gyroradius of newly ionized lunar species. For example, the upper left inset in Fig. 1 shows the result of a computation by Yokota and Saito (2005) of the expected differential fluxes of Na at ~100 km altitudes resulting from both surface sputtering and exospheric photoionization. The surface ions are nearly monoenergetic, while the atmospheric ions spread out in energy due to the extended source and low observation altitude. At higher altitudes, the ions would be accelerated still further, resulting in a much more monoenergetic spectrum and much higher ion energies for both surface and exospheric sources (with little reduction in flux unless significant scattering occurs in one ion gyroperiod). At distances within two gyroradii from the source, one may discriminate between ions with different masses by comparing observed energies and gyrophases with those predicted for the cyclical motion shown in Fig. 1, though distinguishing between surface and exospheric ions becomes more difficult.

Kaguya findings have recently advanced our understanding of the solar wind's interaction with the lunar surface as well as the near-Moon wake. Kaguya's in situ ion mass spectrometry was used to confirm the presence of Na^+ in addition to He^+, C^+, O^+, K^+, and Ar^+ in the sunlit lunar exosphere at altitudes ~100 km when the Moon was in the solar wind (Yokota et al. 2009) and Earth's magnetosphere (Tanaka et al. 2009). Some ions originated from the exosphere, others from the surface. Other species may be present as well. The heavy ion flux varies with solar zenith angle but not with solar wind flux or meteor shower occurrences, suggesting a stable driver for the sputtering process.

Kaguya observations also indicate that from 0.1 to 1% of the solar wind protons incident upon the Moon are reflected. Solar wind convection electric fields accelerate some of these protons to speeds triple those of the solar wind (Saito et al. 2008). The pickup ions find access not only to the high latitude wake, but also areas deep within the low latitude and low altitude wake through fully kinetic processes. Knowledge of the solar wind electric and magnetic fields suffices to reconstruct observed proton spectra from dayside reflected proton sources.

The ESA and SST instruments on ARTEMIS will measure the energy and direction of pickup ion particle distributions, while the EFI and FGM instruments will measure the ambient electric and magnetic fields. With this information, researchers will be able to backtrack observed pickup ions to either surface or exospheric sources. Furthermore, they will be able to roughly determine the ion mass, since both the ion energy and the size of the cycloidal trajectories scale with mass. ARTEMIS can therefore use pickup ion measurements to remotely probe the properties of the surface and the exosphere. Measurements from the ARTEMIS probe measuring the lunar exosphere can then be combined with observations from the other ARTEMIS probe to determine the response of the exosphere to solar wind drivers.

Although ARTEMIS will generally be further away from the Moon than Kaguya, the large geometric factor of the ESA total ion instruments (a factor of ~10 greater than those on Kaguya) will enable sensitive measurements of the pickup ions under stable solar wind conditions over distances ranging from 100 km to several thousand km from the lunar surface. When applied as a function of lunar phase the technique will determine the dependence of the lunar exosphere on lunar longitude undergoing illumination, thereby providing the ion composition versus selenographic longitude. Since the pickup ion trajectories measured by ARTEMIS can be back-traced, the latitudinal exospheric ion distribution can also be determined.

A second method of indirect detection of lunar ions may also be possible using ARTEMIS measurements. Pickup ions form ring distributions that can generate electromagnetic ion cyclotron waves. It may be possible to detect these waves using the ARTEMIS vector magnetic field data. Waves from pickup ions have been identified in the plasma environments of many solar system bodies, including Venus, Mars, and even the moons of the giant planets (Russell et al. 1990; Kivelson et al. 1996; Paterson et al. 1999; Grebowsky et al. 2004; Dougherty et al. 2006; Russell and Blanco-Cano 2007; Delva et al. 2008), and their presence or absence can be used to constrain the local density of pickup ions, and therefore the source strength.

Finally, synergistic measurements made by ARTEMIS and other orbiting spacecraft can be used to constrain exospheric properties. The Lunar Reconnaissance Orbiter (LRO) mission carries the Lyman-Alpha Mapping Project (LAMP) ultraviolet instrument (Gladstone et al. 2010), while the upcoming Lunar Atmosphere Dust Environment Explorer (LADEE) mission carries the NMS for neutral mass spectrometry and the UVS ultraviolet/visual spectrometer (Delory et al. 2009). These missions thus measure gases before ionization, while ARTEMIS measures them post-ionization. By coordinating ARTEMIS measurements with those from these and other spacecraft, we can greatly advance our understanding of the lunar exosphere and its coupling to the surface and the space environment.

2.2 Surface Charging, Electric Fields, and Dust

Although one might think of the lunar environment as essentially dormant, it is in fact very active electrically. Since the Moon has only a tenuous exosphere and no global magnetic field, its surface lies directly exposed to the impact of solar UV and X-rays as well as solar wind plasma and energetic particles. This creates a complex and dynamic lunar electric environment, with the surface typically charging positive in sunlight and negative in shadow to potentials that vary over orders of magnitude in response to changing solar illumination and plasma conditions. These potentials have been measured by instruments on the surface and by Lunar Prospector in orbit, but we still do not fully understand the near-surface environment, especially the role of secondary and photoemitted electrons, and the structure of the plasma sheath at the surface.

The largest observed lunar potentials typically occur on the nightside, in the absence of photoemission, where ambient plasma currents primarily drive surface charging (Manka 1973; Stubbs et al. 2007). Using measurements of the angular distribution of reflected electrons and accelerated secondary electrons from the surface, Lunar Prospector provided estimates of negative nightside lunar surface potentials of ~ -100 V or less in the wake and magnetospheric tail lobes (Halekas et al. 2002a, 2008c), and occasionally as high as -2–4 kV in the magnetospheric plasmasheet (Halekas et al. 2005, 2008c) and during SEP events (Halekas et al. 2007). However, all Lunar Prospector measurements were fundamentally handicapped by a lack of spacecraft potential measurements and the spectrometer's rather coarse electron energy resolution. ARTEMIS's high resolution full plasma and electric field measurements will enable researchers to measure lunar surface potentials from orbit accurately and precisely, as well as determine their response to external influences. The excellent energy resolution of the electron spectrometers on ARTEMIS will also enable a better understanding of the importance of lunar photoelectrons and secondary electrons, and may allow us to determine whether the potential in the plasma sheath varies monotonically, or non-monotonically as suggested by recent simulations (Poppe and Horanyi 2010).

Near-surface electric fields like those described above can loft micron-sized charged dust grains (Nitter et al. 1998). The horizon 'glow' observed by cameras on the landed Surveyors

5, 6 and 7 at lunar twilight was attributed to sunlight scattering by dust electrostatically-levitated just above the lunar surface by electric fields at the terminators (Criswell 1973; Rennilson and Criswell 1974). Astronauts in Apollo's command module also witnessed a twilight horizon glow extending 10's of kilometers above the lunar surface which was suggested to result from surface ejection of submicron grains to high altitudes (McCoy and Criswell 1974; McCoy 1976). Lunar Prospector has confirmed the presence of strong and dynamic near-surface electric fields that are capable of lofting small charged grains to high altitudes (10's of km) in a dust 'fountain' effect (Stubbs et al. 2006). These near surface electric fields may be strongest at the terminator—the starting location of the lunar wake (Farrell et al. 2007).

The nature of the wake plasma discontinuity at the terminator remains unknown. Ideally, there should be a 'perfect' plasma discontinuity at the terminator dividing flowing solar wind plasma from a perfect plasma void. Due to their higher thermal velocities, electrons should migrate into the plasma void further than ions, thereby creating a very large ambipolar E-field just above the lunar surface. This initial plasma expansion E-field could drive dust transport at the terminator (Farrell et al. 2007). However, plasma discontinuities tend to dissipate by radiating plasma waves at group velocities that exceed the speed of the discontinuity. These oscillatory waves and their related ULF/VLF electric fields may diminish the ambipolar effect and associated dust lofting.

ARTEMIS will examine the wake flank/discontinuity as a function of distance from the terminator, providing enough passes under varying solar wind and IMF directions to provide a unique evaluation of the effectiveness of the ambipolar forces in surface dust lofting. In essence, ARTEMIS will cross the discontinuity at various stages of its evolution, providing detailed observations of the plasma expansion and discontinuity dissipation processes. For similar solar wind and IMF configurations, the wake flank can be considered time stationary, and passages through the flank at various altitudes are equivalent to examining the discontinuity dissipation process as a function of time. The initial plasma expansion process is not only fundamentally new science, but also crucial to understanding surface dust lofting from E-fields at the terminator.

2.3 Internal Structure

Electromagnetic sounding encompasses a wide variety of inductive methods used to sense the interior of the Earth, the Moon and other planetary bodies (Schuster and Lamb 1889; Grant and West 1965; Sonett 1982; Khurana et al. 1998; Simpson and Bahr 2005). Suitable signals for planetary-scale soundings arise from the solar wind, magnetospheres, ionospheres, and lightning, depending on the planetary environment (Grimm 2009). These time-varying electromagnetic source fields induce eddy currents in planetary interiors, whose secondary electromagnetic fields are detected at the surface. The secondary fields shield the interior according to the skin-depth effect, which is exploited by electromagnetic sounding by using impedance measurements over a range of frequency to reconstruct conductivity over a range of depth.

Electromagnetic induction studies performed during Apollo gave an indication of the deep electrical conductivity profile and limited the radius of any strongly conducting core to less than about 20% of the lunar radius (e.g, Sonett et al. 1972; Dyal et al. 1974; Hood et al. 1982; Hood and Sonett 1982). Laboratory conductivity-temperature measurements were used to translate the lunar conductivity profile to a selenotherm that is roughly equivalent with the background, global heat flow derived by Wieczorek et al. (2006). The outer 500 km was poorly constrained due to the limited magnetometer bandwidth (higher frequencies are needed to resolve shallower depths).

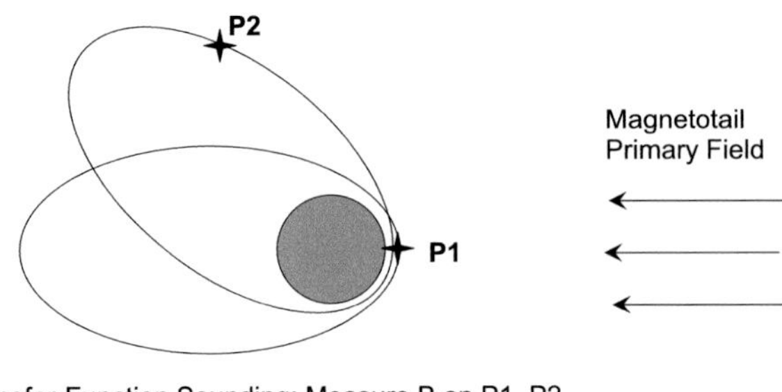

Fig. 2 The ARTEMIS spacecraft probe the lunar interior individually by the magnetotelluric method and jointly using the magnetic transfer function method. The investigation will provide constraints on core size, on the composition of the lower mantle, and on the nature of upper-mantle heterogeneity, particularly the anomalous Procellarum KREEP Terrane

In spite of these advances, significant questions remain. The International Lunar Network Science Definition Team Report (1998) identified three new investigations for electromagnetic sounding: (1) determine the structure of the outer 500 km (upper mantle and crust) that may contain key information on early lunar evolution, specifically, the extent of the magma ocean and the origin of major provinces; (2) provide tighter bounds on the conductivity of the lower mantle (500–1400 km depth), that may constrain trace amounts of H_2O (Saal et al. 2008; Grimm and McSween 2009); and (3) determine whether the Moon has a core, and is it metal or molten silicate.

The fundamental quantity that must be measured in any induction experiment is the frequency dependent electromagnetic impedance Z. The most important constraint is that two known quantities are always necessary (Grimm 2009), e.g., Ohm's Law $Z = V/I$. Apollo-era lunar studies used the distantly-orbiting Explorer 35 to determine the source magnetic field, and compared this to Apollo 12 surface measurements that represent the sum of source and induced fields. This is the *Transfer-Function* technique (see Fig. 2). Alternatively the *Magnetotelluric Method* can provide a complete sounding from individual platforms using orthogonal horizontal components of the electric and magnetic fields (e.g., Vozoff 1991; Simpson and Bahr 2005; Grimm and Delory 2008). The plane-wave impedances determined by the Magnetotelluric Method are extendable to spherical geometry when skin depths become comparable to the planetary radius (Weidelt 1972; see also Schubert and Schwartz 1972).

With these considerations, we have defined an electromagnetic sounding investigation for ARTEMIS. Transfer Function soundings can be performed when the Moon is in the Earth's magnetotail, both probes are in daylight, one probe is near periapsis ($\sim$100 km), and the other at least several lunar radii away (including up to apoapsis $\sim$19,000 km). This ensures that the distant probe measures only the primary field while the periapsis probe measures the sum of the primary and secondary fields. Only the fluxgate magnetometers are necessary. This geometry contrasts with the free-solar-wind environment used for Apollo Transfer Function soundings. Here, the dayside confining plasma layer would have required the second measurement (primary+secondary field) to be made at or near the surface, well below the ARTEMIS periapsis altitude. Measurements with both spacecraft in the wake are also possible, although it may be more difficult to separate the relevant signals in the hot tenuous plasma in this shadowed region, where spacecraft charging may reach extremes.

The probe periapses of 100 km restrict the minimum investigation depth to a few hundred km or greater. They may glimpse some upper mantle structure but will certainly probe the lower mantle, especially given that measurements may be made during some tens of minutes around periapsis. Close study of field-line deflection over many periapses passages may be used to detect the core (Russell et al. 1982; Hood et al. 1999).

ARTEMIS will conduct the first extraterrestrial tests of the Magnetotelluric Method, using both electric (EFI) and magnetic (FGM) records during periapsis passages. While these measurements may be possible under a variety of conditions and geometries, they are likely best conducted in sunlight while in the magnetotail. As these are the same conditions favoring Transfer Function soundings, results should be identical: discrepancies will highlight the kinds of data processing and/or auxiliary information needed to extract the electromagnetic part of the electric-field.

With its 1.5-yr mission at low periapses, ARTEMIS will have both data redundancy (signal integration) and geographic coverage around the equator. The ARTEMIS mission will advance our understanding of the deep lunar interior and provide a baseline for next-generation lunar and planetary electromagnetic sounding.

2.4 Magnetic Anomaly Interactions

Maps for lunar crustal magnetic fields remain significantly undersampled and the causes of lunar crustal magnetization remain uncertain. Local enhancements in the crustal magnetic field can shield the lunar surface from space weathering caused by ion sputtering and generate mini-magnetospheres whose effects can be felt far from the lunar surface. ARTEMIS will provide the combination of in situ and remote observations needed to substantially improve our global picture of crustal magnetism and then determine its effects on its immediate environment.

2.4.1 Causes of Crustal Anomalies

Patterns of crustal magnetization preserve a record of planetary evolution. Thermoremanent magnetization in the presence of a dynamo magnetic field is the dominant process by which the terrestrial crust has been magnetized, as is likely the case for Mars (Connerney et al. 1999). Alternating ‘stripes’ of magnetized seafloor crust at mid-ocean ridges were crucial in confirming geodynamo polarity reversals and the plate tectonics hypothesis (Vine and Matthews 1963; Vine and Wilson 1965). The origin(s) of lunar magnetism are far less certain: lunar sample measurements indicate the possible existence of a lunar dynamo from 3.6–3.9 billion years ago (Cisowski et al. 1983) with an order of magnitude decrease before and after that period.

There are many hypotheses for the source of the ambient field and the magnetization process. Steady magnetizing fields of both external (solar or terrestrial) and internal (lunar dynamo) origin have been proposed, as have transient fields generated by impacts (for reviews, see Fuller 1974, and Wieczorek et al. 2006). Thermoremanent magnetization is likely the dominant process for igneous lunar samples; however, shock remanent magnetization may be significant in lightly metamorphosed breccias, which carry the strongest and most stable remanent magnetization of all lunar samples because they contain more metallic iron grains (likely produced by meteoritic impacts) (Fuller et al. 1974; Fuller and Cisowski 1987). Mare basalts, on the other hand, contain less nanophase iron and generally have weaker remanent magnetization (Coleman et al. 1972).

With samples from only a handful of lunar landing sites, orbital magnetic measurements have been relied upon to study the global pattern of crustal magnetism. Two types of data have been used: magnetometer and electron reflection measurements. Magnetometer data provide vector information but are limited by spacecraft altitude and the fact that lunar crustal magnetic fields at satellite altitudes are typically very weak: $\sim$2 and $\sim$3 orders of magnitude weaker than the terrestrial and Martian cases respectively (Connerney et al. 2004). Electron reflection measurements invoke magnetic mirroring and use magnetometer observations together with electron energy and angular distributions measured from orbit to estimate the magnetic field magnitude at the lunar surface (Anderson et al. 1975; Lin et al. 1976). Electron reflection data have the virtue of being particularly sensitive to the weakest crustal fields (Halekas 2003; Mitchell et al. 2008).

Electron reflection and magnetometer measurements from the Apollo 15 and 16 subsatellites (Hood et al. 1981; Lin et al. 1988) and the Lunar Prospector spacecraft (Hood et al. 2001; Halekas et al. 2001; Mitchell et al. 2008) have enabled global mapping of lunar crustal fields. However, the map is severely undersampled, with significant noise and pixelization clearly visible (see Fig. 3), accounting for much uncertainty in geophysical interpretation. ARTEMIS ESA, with better energy resolution than Lunar Prospector or Apollo, will, along with ARTEMIS magnetometer measurements, enable more accurate correction for the effects of electrostatic potentials between the lunar surface and the spacecraft (Halekas et al. 2002b; Mitchell et al. 2008) and therefore provide more accurate crustal field estimates. The two ARTEMIS probes will make many hundreds of low altitude (<400 km) electron reflection measurements within 10° (20° if optimized) of the equator on every periselene pass, resulting in better spatial coverage and higher fidelity crustal field maps in these regions than currently exist.

Existing maps show weaker fields over the mare basalts compared with the highlands (in agreement with the aforementioned sample studies) and reveal that the largest area of strong crustal fields lie in regions diametrically opposed to the Imbrium, Serenitatis, Crisium and Orientale impact basins (see Fig. 3). This led to the hypothesis that crustal magnetization is associated with basin-forming impacts. According to this hypothesis, the hypervelocity (>10 km/s) impacts that form such large basins produce a plasma cloud that expands around the Moon, compressing and amplifying any pre-existing ambient magnetic field at the antipodal point (Hood and Huang 1991), where the focusing of seismic energy and impact of basin ejecta may result in substantial shock remanent magnetization (Hood and Artemieva 2008). A key element of this hypothesis is that it does not require a global dynamo field, although recently Wieczorek and Weiss (2010) have shown that the simplest explanation for some of the strongest anomalies like Reiner-Gamma involves a steady, dynamo magnetic field. Better maps enabled by ARTEMIS in the equatorial regions will help to elucidate the mechanism of antipodal magnetization by allowing us to better constrain the characteristics of the remanent crustal remanent magnetization.

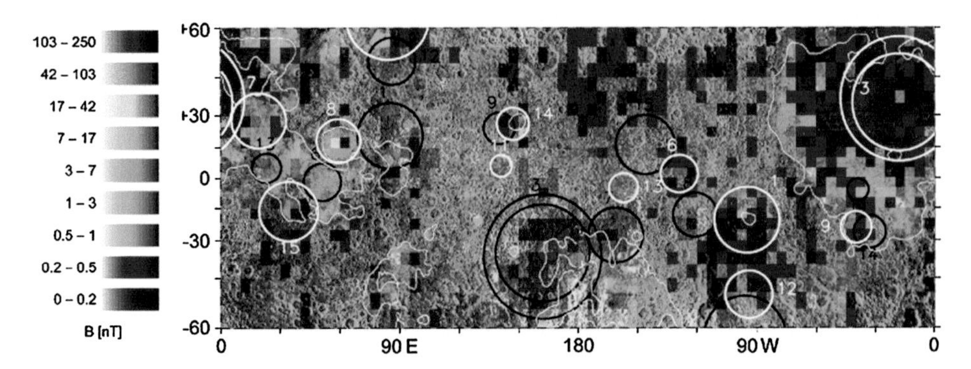

Fig. 3 Impact basin rims (*white circles*, for Imbrium the interior ring is also shown) and their antipodes (*black circles*), superimposed on a map of total surface magnetic field strength averaged over $5 \times 5°$ bins (Mitchell et al. 2008)

In summary, the ARTEMIS electron reflection data set will substantially augment existing observations, enabling greater resolution in the equatorial regions and thus more confident interpretation. In so doing, ARTEMIS will improve our knowledge of the interconnected histories of lunar impacts, basin formation, magmatism and the ancient lunar dynamo.

2.4.2 *Effects of Crustal Anomalies on the Plasma Environment*

The sub-ion inertial scale lengths associated with crustal magnetic anomalies raise a host of questions about the nature of their interaction with the solar wind. Does the solar wind-anomaly interaction produce a shock? Do the magnetic anomalies shield a region from the solar wind, generating a lunar "mini-magnetosphere" that stands off and deflects the solar wind (Hood and Schubert 1980; Hood and Williams 1989; Hood 1992; Lin et al. 1998; Harnett and Winglee 2000)? Past and current missions have led to great progress on these questions, but ARTEMIS stands ready to add to our knowledge, thanks to its comprehensive plasma instrumentation and two-probe design. ARTEMIS offers the unprecedented (at the Moon) capability to measure full 3-D plasma distributions and electric fields, allowing better understanding of the microphysics of the interaction, and to measure the properties of the plasma just upstream from the anomaly region, allowing separation of spatial and temporal effects. The results may be applicable to other bodies, including Mars (e.g., Breus et al. 2005), asteroids such as Gaspra and Ida (Kivelson et al. 1995), and possibly Vesta (Vernazza et al. 2006).

Explorer 35 and the Apollo subsatellites first observed the effects of magnetic anomaly interactions near the limb in the form of compressional waves propagating downstream. Lunar Prospector, however, observed the effects extending to $\sim$100 km altitudes, sometimes well upstream ($>45^\circ$) from the limb and their apparent sources, seemingly requiring the formation of a shock (Lin et al. 1998). The coincident enhancements in the upstream electron density, electron flux, magnetic field strength, and whistler mode wave activity upstream seen in Fig. 4 all suggest the presence of a shock (Lin et al. 1998; Halekas et al. 2006b, 2007, 2008b). By contrast, simulations for the solar wind interaction with dipoles with dimensions comparable to lunar anomalies only predict whistler or magnetosonic wakes, not shocks (Omidi et al. 2002). Some simulations suggest that the non-dipolar nature of lunar magnetic sources might increase the efficiency of the interaction (Harnett and Winglee 2003), perhaps resolving this discrepancy. Also, surface influences, including secondary electrons, photoelectrons, and dust, may turn out to play a role. To determine whether lunar anomalies do in fact form shocks (Halekas et al. 2006a, 2006b), we need the capability of ARTEMIS plasma instrumentation, which thanks to the spinning spacecraft and spacecraft potential measurements, can make precise measurements of bulk properties and anisotropies of the 3-D velocity distributions of both electrons and ions.

2.4.3 *Space Weathering: Effects of Crustal Anomalies on the Lunar Surface*

The Moon, like any other body exposed to the harsh space environment, is subject to galactic and solar cosmic rays, irradiation, implantation, and sputtering from solar wind particles, and bombardment by meteorites and micrometeorites. This exposure causes radiation damage, chemical changes, optical changes, erosional sputtering, and heating, all essential parts of the process called space weathering. Space weathering is important because these processes affect the physical and optical properties of the surface of many planetary bodies. To properly interpret remote sensing observations, it is critical to understand the effects of space weathering.

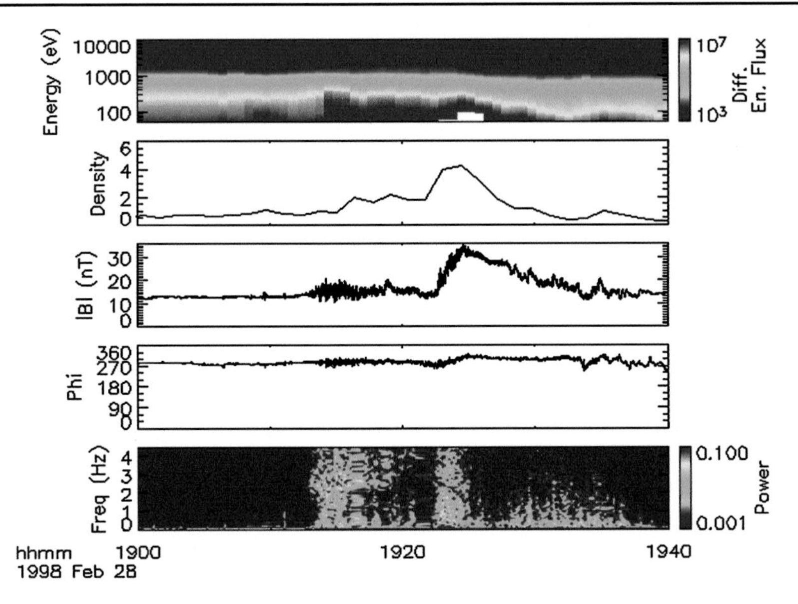

Fig. 4 Lunar Prospector electron energy spectra, electron density, and magnetic field strength, longitude, and power frequency spectra suggestive of a shock

A number of observations suggest that magnetic anomalies can shield the surface from the solar wind, with clear implications for surface weathering, and possible implications for the formation of the lunar albedo markings often observed near strong crustal magnetic sources (e.g., Hood and Schubert 1980; Richmond et al. 2005). Lunar Prospector observations of low altitude density cavities above a strong crustal anomaly lend some support to this idea (Halekas et al. 2008a). More recent Kaguya observations of ion flow deceleration, electron heating, and no scattered protons above strong crustal anomalies also suggest surface shielding (Saito et al. 2010). Finally, Chandrayaan has provided observations indicating no scattered neutral atoms from the same region discussed by Halekas et al. (2008a), strongly suggesting the existence of a magnetically shielded surface region (Wieser et al. 2010).

The two ARTEMIS probes will enable more observations like those by Kaguya, but with more comprehensive plasma instrumentation, and the ability to compare observations upstream and outside of the anomaly region with those inside. This will enable separation of temporal and spatial effects, and allow researchers to clearly determine how the anomaly interaction affects the incoming solar wind flow, including whether and how electrons and ions de-couple near anomalies, what mechanisms produce electric fields, and the nature and distribution of particle heating and wave-particle interactions.

3 ARTEMIS Heliophysics Science Objectives

ARTEMIS will address the host of heliospheric science objectives illustrated in Fig. 5. Topics include reconnection, particle acceleration, and turbulence in the magnetotail and interplanetary space, and the structure and evolution of the Moon's plasma wake. This section presents a selection of topics to be considered as a function of location in orbit: addressing successively the Earth's magnetotail, the solar wind, and the lunar wake.

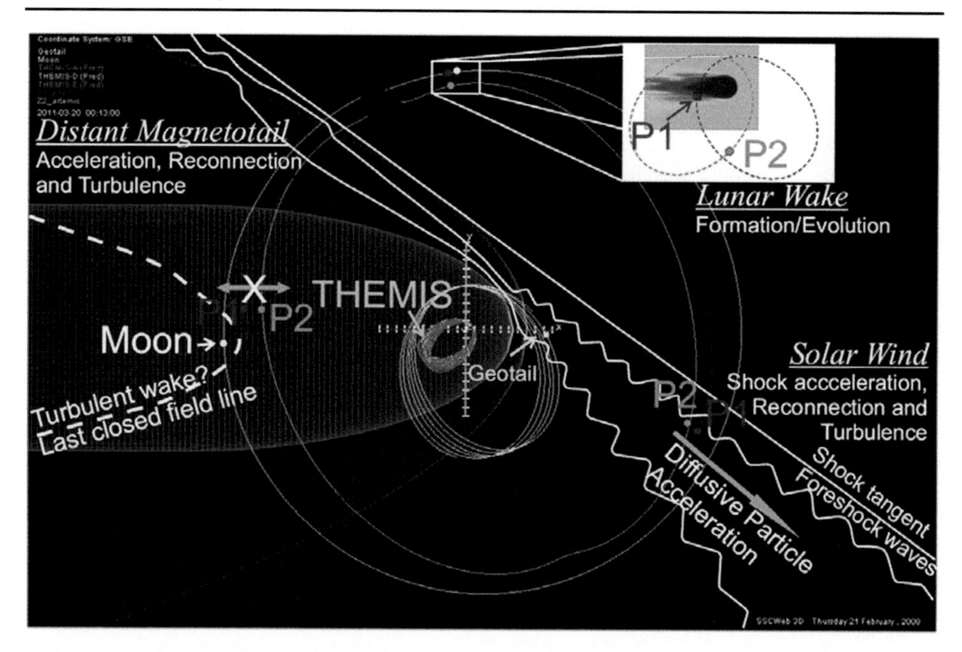

Fig. 5 ARTEMIS heliophysics scientific objectives address topics including reconnection, particle acceleration, and turbulence in the solar wind and Earth's magnetotail and the formation and evolution of the lunar wake

3.1 In the Magnetotail

The Earth's magnetotail comprises two lobes with oppositely directed magnetic fields divided by an equatorial current sheet embedded within a plasma sheet marked by enhanced densities and temperatures but depressed magnetic field strengths. Conditions within the near-Earth magnetotail are relatively well-understood, thanks in large measure to the multipoint measurements returned by the THEMIS mission. By contrast, fleeting glimpses of the distant magnetotail by the Explorer-33 and -35, ISEE-3, Geotail, and Wind missions have raised numerous questions that can only be answered by examining extensive multipoint in situ observations. Potential research topics include the structure of the distant magnetotail during prolonged intervals of northward IMF orientation, the length, shape, and occurrence patterns for reconnection lines at lunar distances, the evolution of the heated and accelerated flows ejected by these reconnection lines, the shape, dimensions, internal structure, and evolution of antisunward-moving plasmoids and flux ropes, boundary waves on the surface of the magnetotail, and the nature of turbulence within the Earth's magnetotail plasma sheet.

ARTEMIS will provide the observations needed to address these questions. Together with the Moon, the two ARTEMIS spacecraft will spend about 4.5 days of each 28-day orbit about the Earth within the magnetotail at lunar distances, and 20 to 30 hours per orbit within the plasma sheet (Hapgood et al. 2007). With interspacecraft separation distances ranging from 500 km to 20 R_E, the two spacecraft will be well situated to determine not only the steady-state macrostructure of the distant magnetotail, but also the meso- and micro-scale characteristics of the superimposed transient events as a function of solar wind conditions.

3.1.1 Structure of the Magnetotail

The first question to be addressed concerns the structure of the distant magnetotail during prolonged intervals of northward IMF. Magnetosheath momentum imparted to the magnetosphere drags magnetospheric magnetic field lines antisunward to form the Earth's magnetotail. During periods of southward IMF orientation, this momentum is imparted via the reconnection of magnetospheric and interplanetary magnetic field lines on the dayside magnetopause. Theory predicts, and observations confirm, that the Earth's polar caps lie on open magnetic field lines leading to interplanetary space during intervals of southward IMF orientation (Dungey 1961). The situation during periods of northward IMF (NBZ) remains unclear. Reconnection on the dayside equatorial magnetopause ceases, terminating the production of open magnetic field lines. Reconnection within the magnetotail continues, transforming open lobe into closed plasma sheet magnetic field lines. The likelihood of dual lobe reconnection, which also closes open lobe field lines while appending magnetosheath magnetic field lines to the magnetosphere, increases. The magnetotail may become topologically closed during prolonged NBZ periods.

Observations from low altitude polar orbiting spacecraft like DMSP can be used to infer the polar cap size, the amount of open flux in the lobes, and whether or not the magnetotail ever closes. Low-altitude observations indicate that the size of the open magnetic field line region within the polar cap slowly diminishes during periods of strong and persistent northward IMF (Newell et al. 1997). However, spacecraft may fail to enter the polar caps when their size diminishes, resulting in a mistaken interpretation that they are absent. Furthermore, their identification can become more difficult during extended NBZ periods because the characteristics of the precipitating particles used to identify their boundaries may change. In the absence of clear predictions for the shape and geometry of the magnetotail and its plasma and magnetic field characteristics during NBZ periods, in situ observations of the magnetotail have also generally been inconclusive. A chance Wind encounter with a strongly deformed and twisted magnetotail some 125 R_E from Earth during an extended NBZ period on October 22–24, 2003 indicates an open magnetotail; albeit one with highly unusual plasma and magnetic field properties (Øieroset et al. 2008).

Global numerical models for the magnetosphere predict strikingly different magnetospheric topologies during NBZ periods. Many predict magnetospheres that close near Earth, as envisioned by Axford and Hines (1961). Others predict closed magnetospheres that extend much further, perhaps in response to viscous momentum transfer enhanced by the Kelvin-Helmbholtz instability (Miura 1984). Predicted lengths for the closed magnetotails vary from as little as $\sim$30 to several hundred R_E (Usadi et al. 1993; Watanabe and Sato 1990; Ogino et al. 1992; Fedder and Lyon 1995; Gombosi et al. 1998; Guzdar et al. 2001). However, not all models predict tail closure during NBZ periods. Raeder et al. (1995) presented idealized simulations of a NBZ period where the tail stayed open for several hours. In fact, Raeder (1999) argued that the prediction of a closed tail in most numerical MHD models was due to excessive numerical diffusion, a point discussed in further detail by Gombosi et al. (2000) and Raeder (2000).

Figures 6 and 7 present Open Geospace General Circulation Model (OpenGGCM) predictions for the size and shape of the Earth's magnetotail during the October 22–24, 2003 Wind event. The width of the magnetotail diminishes with distance downstream, but an open magnetotail that flaps dawnward and duskward is still present at lunar distances. Torques applied by the IMF twist the magnetotail, causing the southern lobe to shift northward at dusk and the northern lobe to shift southward at dawn (e.g., Sibeck et al. 1985)

ARTEMIS will provide the observations needed to determine the size, shape, and internal structure of the Earth's magnetotail as a function of independently measured solar

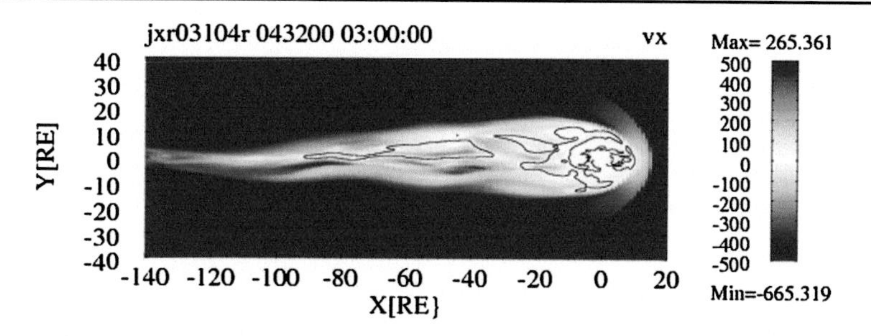

Fig. 6 OpenGGCM simulation predictions for the component of the plasma velocity along the Sun-Earth line during the October 22–24, 2003 strong NBZ event. Magnetotail widths diminish steadily with distance antisunward

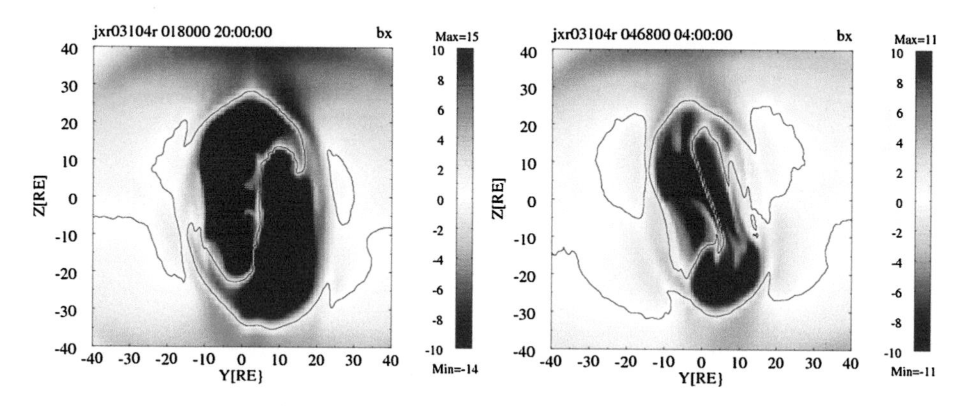

Fig. 7 OpenGGCM predictions for sunward (*red*) and antisunward (*blue*) magnetic field polarities within magnetotail cross-sections 60 R_E from Earth at 20:00 UT on October 22, 2003 and 04:00 UT on October 23, 2003. The magnetotail is severely twisted and deformed

wind and geomagnetic conditions, during both northward and southward IMF orientation. Two point ARTEMIS measurements will aid in determining the orientations of the various boundaries and the characteristics of their flapping motions. One of the two ARTEMIS probes will frequently monitor magnetosheath conditions immediately outside the distant magnetotail magnetopause while the other measures (e.g.) electron pitch angle distributions at the magnetopause and deeper within the magnetotail, the key information needed to reveal magnetic field line topology and history in those regions (Hasegawa et al. 2005; Øieroset et al. 2008). Multi-season and multi-year observations will permit researchers to test whether dual lobe reconnection for northward IMF orientations becomes more efficient when the Earth's dipolar tilt diminishes (Li et al. 2008).

3.1.2 Plasmoids and Flux Ropes: Reconnection Within the Near-Earth Magnetotail

As illustrated in Fig. 8, bursts of reconnection along extended lines in the near-Earth magnetotail release immense antisunward-moving flux ropes, or plasmoids, that carry vast quantities of plasma and magnetic flux antisunward through the Earth's plasma sheet. Spacecraft in the vicinity of the plasma sheet may directly enter the plasmoids, while spacecraft within the magnetotail lobes record perturbation signatures known as traveling compression regions (TCRs). In conjunction with Geotail, Cluster, and baseline THEMIS observations of

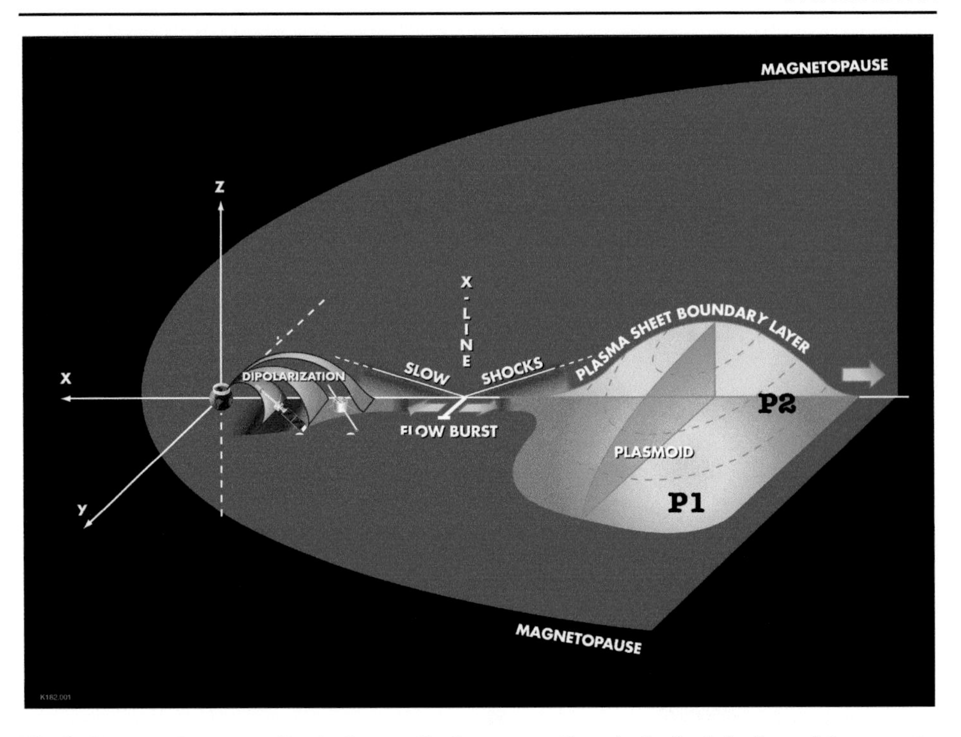

Fig. 8 Reconnection at an x-line in the near-Earth magnetotail results in dipolarizations of the magnetospheric magnetic field and releases enormous plasmoids that move antisunward through the magnetotail. When separated transverse to the axis of the magnetotail, ARTEMIS probes P1 and P2 will be well-situated to determine the dimensions, shape, and internal structure of the plasmoids

the near-Earth magnetotail, and employing its own observations separated by 1–10 R_E along the Sun-Earth line, ARTEMIS will provide the observations needed to determine the internal structure of plasmoids and track their evolution down the magnetotail.

Plasmoid dimensions are a crucial factor in determining their significance to the solar wind-magnetosphere interaction. Plasmoid diameters increase by a factor of 2–3 from near-tail to lunar distances (Ieda et al. 1998), but the reasons for this growth remain unclear. Part of the growth results from expansion in response to lower ambient pressures in the distant magnetotail, part may be due to continued reconnection at reconnection lines moving down the magnetotail with the plasmoids, and part may be due to the coalescence of smaller plasmoids to form larger plasmoids as they are transported tailward. The cross-tail extent of plasmoids remains unknown. Although plasmoids in the distant magnetotail often exhibit force-free structures, their magnetic topology remains uncertain. Magnetic field lines within plasmoids may be closed loops, connected to Earth, or connected to the interplanetary medium.

Azimuthal probe separations of 1–10 R_E will enable ARTEMIS to determine the cross-tail extent, orientation, and shape (using minimum variance analyses of the magnetic field) of plasmoids. Radial separations will provide the observations needed to determine how rapidly plasmoid diameters change with distance downstream (e.g., Slavin et al. 1999). Grad-Shafranov reconstruction techniques (Hasegawa et al. 2004) will be used to determine the plasma and magnetic field structure within the plasmoids from time series measurements and an assumption that they are in near pressure balance with their immediate environment.

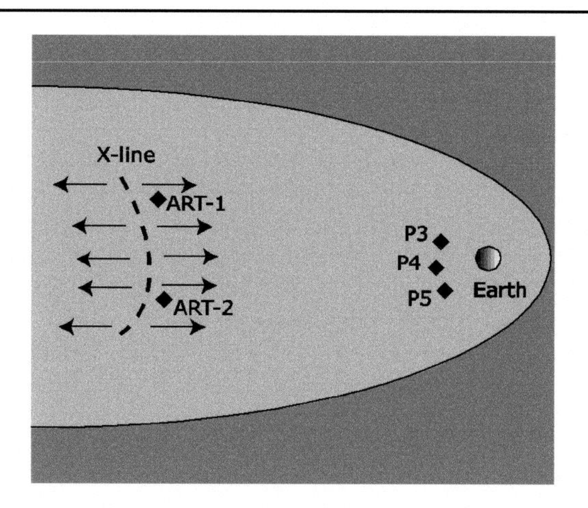

Fig. 9 Reconnection along an extended and curved line in the distant magnetotail. Because the line demarcates the boundary between sunward (Earthward) and antisunward flows, its location can be determined when ARTEMIS probes P1 and P2 are radially separated. By contrast, its extent can be determined when the two spacecraft lie separated transverse to the axis of the magnetotail

High time resolution ARTEMIS measurements of multiple pitch angle distributions during events with anticipated time durations ranging from 100 to 200 s will not only offer an opportunity to determine their magnetic field topology, but also to test predictions concerning reconnection-related electron energization in contracting magnetic islands (Drake et al. 2006).

3.1.3 Occurrence Patterns for Reconnection in the Distant Magnetotail

By contrast to the situation in the near-Earth plasma sheet, where reconnection associated with geomagnetic substorms is often bursty and patchy (Baumjohann et al. 1989, 1990; Angelopoulos et al. 1992), ISEE-3, Geotail, and Wind observations of long-duration reconnection-heated plasma jets suggest that reconnection in the distant magnetotail plasma sheet is often quasi-steady and large-scale. As illustrated in Fig. 9, reconnection may occur along curved reconnection lines whose cross-tail extent depends upon the IMF orientation (Feldman et al. 1985; Nishida et al. 1995; Angelopoulos et al. 1996; Øieroset et al. 2000). In the absence of multi-point measurements, we do not know either the conditions favoring localized and extended reconnection in the distant magnetotail or the shape of the reconnection line. Two-point ARTEMIS observations enable both the occurrence patterns of reconnection and its extent across the lunar magnetotail to be determined as a function of guide field strength, geomagnetic, and solar wind conditions.

3.1.4 Heating, Particle Acceleration, and Plasma Transport at Distant Reconnection Lines

The plasma sheet within the distant magnetotail is an excellent location to study magnetic reconnection. By contrast to the near-Earth magnetotail, there is no heating in response to flow braking against the dipolar magnetic field lines. Furthermore, the finite mantle densities in the distant magnetotail mean that plasma parameters in the inflow region can be better characterized than they would be in the very tenuous near-Earth magnetotail. Topics for investigation by ARTEMIS include the process(es) by which reconnection heats plasmas, particle acceleration in reconnection geometries, and the subsequent motion of the reconnected plasma.

The degree to which reconnection heats ions, and the factors controlling this heating, are presently not known. Results from some recent simulations suggest that the heating is proportional to the speed of the outflowing Alfvénic jets, which are in turn related to the

Alfvénic velocities of the inflowing plasma (Drake et al. 2009). The two ARTEMIS spacecraft will provide numerous opportunities to simultaneously compare inflow and outflow parameters.

Particles can be accelerated to suprathermal energies by drifting along X-lines, but also by Fermi-acceleration in the collapsing bubbles that surround O-lines (Hoshino et al. 2001; Drake et al. 2006). Wind observations of a single event in the distant magnetotail (at $X_{\mathrm{GSE}} = -60\ R_{\mathrm{E}}$) provide evidence for electron energization to at least 300 keV (Øieroset et al. 2002). Two-spacecraft ARTEMIS observations of plasma flows and magnetic field components normal to the current sheet are essential to discriminate between (and track the motion of) X-lines and O-lines. ARTEMIS will provide the previously unavailable high resolution plasma and electric field measurements needed to determine the causes for electron energization.

Steady reconnection in the distant magnetotail should eject high-speed plasma flows and energized particles both Earthward and away from the Earth. The occurrence patterns and extents of distant magnetotail reconnection lines will be determined from two-point ARTEMIS observations of the Earthward (antisunward) streaming energetic particles in the plasma sheet boundary layer, sunward (antisunward) flows, and northward (southward) magnetic field components in the current sheet expected sunward (antisunward) from the reconnection line. There are indications that most of the sunward plasma jets from the distant tail do not reach the near-Earth plasma sheet (e.g., Øieroset et al. 2004). Working in conjunction with the three inner THEMIS spacecraft, Cluster, and Geotail, ARTEMIS will establish the relationships between plasma flows in the lunar and near-Earth plasma sheet to determine the fate of reconnection jets generated in the distant tail. The results will help distinguish between models in which the flows are decelerated and/or deflected towards the flanks.

3.1.5 Boundary Waves, Flux Transfer Events, and Turbulence in the Magnetotail

The Kelvin-Helmholtz (KH) instability may play an important role in transferring solar wind mass, momentum, and energy into the Earth's magnetotail (Fujimoto and Terasawa 1994; Miura 1984, 1992), perhaps dominating the overall solar wind-magnetosphere interaction during intervals of northward IMF orientation when reconnection shuts down on the dayside magnetopause. The instability is most likely along the low-latitude flank magnetopause, where and when flow shears lie nearly perpendicular to weak northward magnetosheath and magnetospheric magnetic fields (Southwood 1968). As illustrated in Fig. 10, the amplitudes and wavelengths of waves generated by the Kelvin-Helmholtz instability should grow

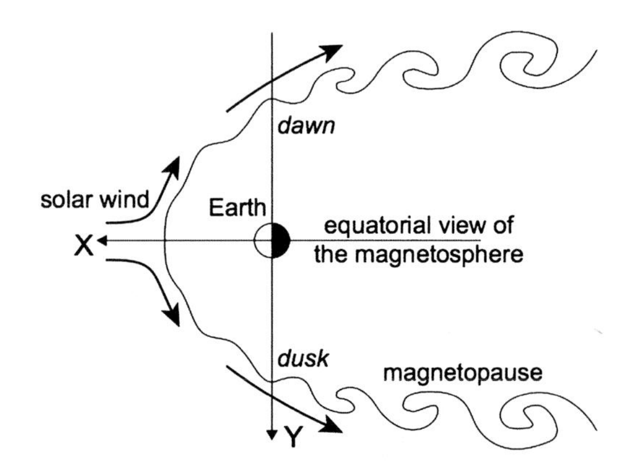

Fig. 10 Illustrating the expected evolution of the Kelvin-Helmholtz instability from a linear stage on the dayside magnetopause to a nonlinear stage on the flanks of the distant magnetotail

with downstream distance. The waves may merge into larger vortices (Miura 1997), perhaps with steepened trailing edges (Chen and Kivelson 1993). They may also generate fast-mode shocklets (Lai and Lyu 2006) or induce turbulence (Matsumoto and Seki 2007). Simulations that include the effects of coupling with the ionosphere and/or the tail lobes indicate that finite plasma sheet thicknesses may suppress the nonlinear growth of the instability (Galinsky and Sonnerup 1994; Takagi et al. 2006).

Observations indicate that the KH instability is relatively common on the flanks of the near-Earth magnetotail, occurring for the predicted solar wind conditions and exhibiting the expected signatures (Fairfield et al. 2000; Hasegawa et al. 2006). The structural properties of the KH waves are similar on the dawn and dusk flanks of the near-Earth magnetosphere (Hasegawa et al. 2006; Nishino et al. 2011). However, non-MHD processes that depend on the polarity of the field-aligned vorticity could lead to dawn-dusk asymmetries further downstream (Nakamura et al. 2010).

Although the distant magnetotail magnetopause is constantly in motion, and this motion has occasionally been attributed to the KH instability (Sibeck et al. 1987), little is known about the properties of the instability at downstream distances beyond 30 R_E. ARTEMIS will provide the simultaneous two-point observations of magnetosheath and magnetosphere needed to test instability criteria and determine the velocities, amplitudes, and wavelengths of the resulting boundary waves. Both the two-point observations, and comparisons with simultaneous Cluster and Geotail traversals of the near-Earth magnetopause, will help determine whether waves breakup, coalesce, expand, stabilize, or simply propagate downstream. The same two point observations will be examined for evidence of KH-generated reconnection, diffusion, turbulence, shocklets, and dawn/dusk asymmetries (e.g. Chaston et al. 2007).

Observations on the dayside and near-Earth flank magnetopause reveal the presence of twisted ropes of interconnected magnetosheath and magnetospheric magnetic field lines marked by transient magnetic field strength increases and bipolar magnetic field signatures normal to the nominal magnetopause. Interpreted as flux transfer events or FTEs, the flux ropes provide evidence for the frequent occurrence of unsteady reconnection at the magnetopause (Russell and Elphic 1978; Raeder 2006). The significance of the events to the overall solar wind-magnetosphere interaction remains to be determined. If entirely disconnected from the ionosphere, the events may simply advect antisunward with the magnetosheath flow, maintaining their orientation and helical magnetic field geometry, and represent nothing more than an interesting curiosity. However, if one end of the flux ropes remains anchored in the ionosphere while the other extends into interplanetary space, the events may account for a significant transfer of solar wind mass, energy and momentum to the magnetosphere. In this case, the flux ropes will stretch out parallel to the Sun-Earth line and sink into the magnetotail lobes (Sibeck and Siscoe 1984). Two-point ARTEMIS measurements will help determine event motion, orientation, length, and topology at lunar distances, thereby providing crucial information concerning their role in the solar wind-magnetosphere interaction.

Turbulent dissipation is an effective mechanism for heating fluids and transferring mass, momentum and energy. Although turbulence is ubiquitous in space plasmas, much remains to be determined concerning its initiation and subsequent evolution. Turbulence in the near-Earth plasma sheet has been studied using Cluster (Weygand et al. 2007). The dissipation range was on the order of the ion inertial lengths or gyroradius ($\sim$few hundred km) and the correlation coefficients diminished to zero beyond scales of 3 R_E. Flow fluctuations were small relative to the sound and Alfvén speeds (except during dynamic conditions). By contrast, flow fluctuations are comparable to the sound and Alfvén speeds in the distant magnetotail and therefore dynamically and energetically important. Theory and simulations point towards magnetotail reconnection and velocity shears at the flanks as likely drivers of plasma sheet turbulence. Both drivers can affect energy circulation and particle transport within the magnetosphere. Characterizing the nature of the fluctuations, and determining their origin and dissipation is therefore important for global circulation. It is quite likely that the distant tail also exhibits an inertial range of turbulence. ARTEMIS will use varying interspacecraft separation distances to characterize turbulence over a wide range of spatial scales.

ARTEMIS may also observe turbulence at the distant magnetotail magnetopause. During periods of northward or nearly radial IMF orientation, reconnection on the high-latitude and flank magnetopause results in high speed antisunward flows in a boundary layer of open magnetic field lines disconnected from Earth. High-latitude reconnection can continue for hours under northward IMF conditions (Frey et al. 2003; Hasegawa et al. 2008). The distant high-latitude magnetopause is an ideal region to study turbulence in reconnection outflows. By contrast to the dayside magnetopause, the effects of Alfvén waves reflected from the ionosphere need not be considered. And by contrast to the near-Earth plasma sheet, the resistance of the dipolar magnetic field region need not be addressed. ARTEMIS will determine whether or not a quasi-steady inertial range of turbulence is attained as a result of energy injection via reconnection. The two-point measurements permit estimations of the wavelength (instead of or in addition to the frequency) of the associated waves (Chaston et al. 2008). A spacecraft in the magnetosheath can provide information about the nature of magnetosheath turbulence and whether conditions favor high-latitude reconnection, thereby helping discriminate between the effects of magnetosheath and reconnection-induced turbulence.

3.2 In the Solar Wind

Together with the Moon, the ARTEMIS spacecraft will spend most of their orbit about the Earth in the solar wind. Here ARTEMIS provides a unique opportunity to address longstanding questions concerning the physics of the foreshock, interplanetary shocks, reconnection in the solar wind, and plasma turbulence. In particular, ARTEMIS will be used to:

- examine how the bow shock, and collisionless shocks in general, accelerate particles to high energies;
- study the structure of interplanetary shocks, and examine how non-planar structure can, for example, influence the production of type II radio emission;
- explore reconnection in the solar wind, and provide the high time resolution measurements required to understand low magnetic shear reconnection, which recent measurements suggest predominates in the heliosphere.
- fill in the gaps in our knowledge of solar wind turbulence by providing cross- observations of solar wind features from spacecraft separated by previously inaccessible interseparation distances of 11,000–50,000 km.

3.2.1 Shock Physics—Particle Acceleration in the Terrestrial Foreshock

In addition to mediating the flow of super-magnetosonic plasma, collisionless shocks also act as sites for particle acceleration (see, e.g., reviews by Terasawa 2003; Burgess 2007). If the upstream magnetic field is not perpendicular to the shock normal, then a portion of the inflowing plasma can escape back into the upstream region rather than being processed by the shock; the interaction of this backstreaming component with the inflowing plasma leads to wave generation and particle acceleration (Eastwood et al. 2005). The way in which collisionless shocks produce energetic particles is a problem of extremely broad astrophysical importance. The Earth's bow shock/foreshock is one of the best laboratories we have for studying in-situ the basic physical processes that govern shock particle acceleration (Burgess 2007).

Diffusive shock acceleration (Axford et al. 1977; Bell 1978a, 1978b; Blandford and Ostriker 1978) is widely cited as the process by which ions are accelerated to high energies at shocks. At the Earth's bow shock, ion energies extend to at least several MeV (Lin et al. 1974; Desai et al. 2000). Diffusive shock acceleration theories predict that the density of energetic ions falls exponentially with distance from the shock front into the upstream region. On a statistical basis, single spacecraft observations just upstream of the bow shock have shown that the energetic ion flux e-folding distance varies from 3.2 +/−0.2 Re at 10 keV to 9.3 +/−1.0 Re at 67 keV (Trattner et al. 1994). In a two-spacecraft case study using Cluster, Kis et al. (2004) and Kronberg et al. (2009) found the e-folding distance varied from 0.5 Re at 11 to 2.8 Re at 27 keV, and increased almost linearly with energy up to ∼120 keV. Although two-point observations provide a far more reliable measurement of the e-folding distance, the Kis et al. (2004) and Kronberg et al. (2009) multipoint analysis studied only one event, thus limiting their conclusions to high Mach number solar wind in a limited location (a few R_E upstream of the shock) over a limited time interval (8 hours of data due to the orbit of the spacecraft). Evidently a multi-point statistical survey is desirable and necessary to better understand the nature of accelerated particles upstream of the bow shock.

ARTEMIS will make extensive observations of the upstream region where particles are accelerated to high energies. ARTEMIS will not cross the sub-solar bow shock except during extreme solar wind conditions, e.g. very low solar wind Mach numbers (Farris and Russell 1994), that would in itself result in serendipitous scientific discoveries. The presence of two spacecraft allows correlation lengths parallel and perpendicular to the field to be studied in the key range of 0.1–20 R_E without the confounding effects of upstream variability—indeed observations can be quantified according to upstream conditions. Foreshock energetic particles will be observed both intermittently and continuously over many days, providing excellent statistics. The ARTEMIS probes will sample a much wider range of distances from the shock than previously accessible, at various distances from the tangent line and over a wide range of depths in the foreshock.

ARTEMIS will also provide experimental data that challenges the common use of linear diffusive shock acceleration theory. At the bow shock there is already incontrovertible evidence that the linear theory is insufficient; for example, magnetic field fluctuations do not exhibit the power law spectrum assumed by models over the required frequency range (Terasawa 1995). Although linear diffusive shock acceleration theory has been extended into the quasi-linear regime (where the energy flows from the particles to the waves, but the waves themselves are given by linear theory) (Lee 1983), experimental tests of quasi-linear theories at the Earth's bow shock have not been carried out (Burgess 2007). Furthermore, recent large-scale hybrid simulations predict that rather than continuing to diminish, the energetic particle flux approaches a constant at some point far upstream of the shock (Giacalone 2004). ARTEMIS measurements will help to establish the homogeneity of the upstream wave field, of key importance for developing more complex theories of particle acceleration in the foreshock.

Finally, as noted by Burgess et al. (2005), diffusive shock acceleration has also often been challenged by the hypothesis that all upstream ion enhancements at Earth are exclusively magnetospheric in origin (e.g. Sarris et al. 1987). Although studies have concluded that most upstream energetic ions do not originate in the magnetosphere (Gosling et al.

1989), the presence of energetic magnetospheric oxygen in the foreshock has been reported (Mobius et al. 1986). The combination of ARTEMIS, making measurements in the foreshock, together with the remaining three THEMIS spacecraft at the magnetopause and in the magnetosphere measuring changes in energetic ions and the state of the magnetosphere, will allow the magnetospheric input of energetic particles to the shock to be deconvolved from diffusive shock acceleration better than ever before for a variety of solar wind conditions.

3.2.2 The Structure of Interplanetary Shocks

Interplanetary (IP) shocks energize particles, and observations in their vicinity can be used to test theories of particle acceleration. They are thought to be responsible for the production of Solar Energetic Particles (SEPs) in so-called gradual events. In one study of particular note, Kennel et al. (1986) used ISEE-3 observations of an IP shock to test the Lee (1983) quasi-linear theory of diffusive shock acceleration. A number of disagreements were observed which could be explained by variations in the upstream conditions, but since data from only a single spacecraft were available, this could not be resolved. By simultaneously observing the conditions both up- and downstream of interplanetary shocks, and then generating statistics based on observations of many IP shocks during the course of the mission, ARTEMIS will provide new information about how IP shocks generate energetic particles. The routine acquisition of spin resolution (3 s) plasma moments will be of particular importance in this regard. Figure 11 shows an example of an IP shock observed by both ARTEMIS spacecraft on 5 April 2010. At this time the spacecraft were in transition from terrestrial to lunar orbit.

Related to this issue is the nature of type II radio emission. Type II radio emissions are thought to be generated patchily in the region upstream from Coronal Mass Ejection (CME)-driven interplanetary shocks (Bale et al. 1999; Pulupa and Bale 2008), and are a useful tool

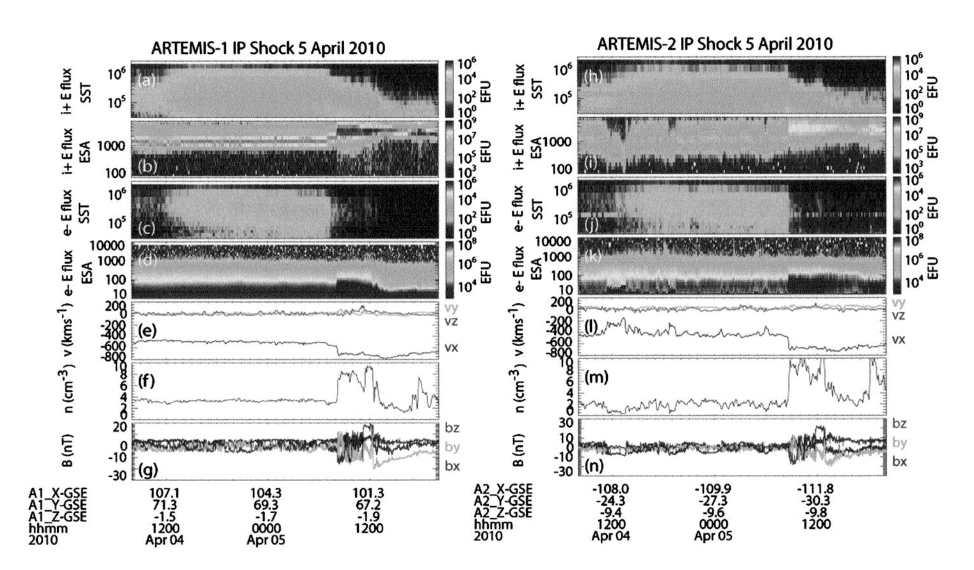

Fig. 11 ARTEMIS P1 (*left*) and P2 (*right*) observations of an interplanetary shock on 5 April 2010. From top to bottom, the panels show: SST ion energy flux, ESA ion energy flux, SST electron energy flux, ESA electron energy flux, ESA ion velocity components, ESA ion density, and FGM magnetic field components. Note that because the spacecraft were in transition from Earth to lunar orbit, they were widely separated at this time. The ARTEMIS-1 encounter occurred at 08:12 UT and the ARTEMIS-2 encounter at 08:42 UT. Both ARTEMIS spacecraft detected significant fluxes of energetic electrons and ions

for tracing the propagation of CMEs through the heliosphere. It is still unclear whether the structuring of type II emissions is due to the curvature of the shock on scales of 10 s of Earth radii, or due to the curvature of the magnetic field, although it is known that upstream turbulence alone cannot explain the dimensions of the acceleration regions inferred from a single spacecraft (Pulupa and Bale 2008). Furthermore, the ultimate cause of any such rippled shock structures remains unclear, especially as a function of upstream conditions. Since the emissions are controlled by the shock structure, to better understand how type II emission is produced (and thus improve its utility in tracking CMEs) it is necessary to study how shock structure varies with upstream plasma beta, Mach number (also related to the CME speed) and magnetic field orientation.

ARTEMIS measurements will allow investigation of IP shock rippling and inhomogeneities on scales of 0.1–20 R_E, highly relevant to the problems discussed here (Pulupa and Bale 2008). Two-point measurements of magnetic field vectors will help determine the curvature of magnetic field lines, while two-point measurements of shock normals will help determine shock curvature. Targeted burst mode operations will allow high (3 s) time resolution 3-D ion and electron distributions to be captured, providing new details about the upstream structure and the presence of electron beams. Burst mode operations can be targeted to capture the interplanetary shock crossing at high time resolution, including the 3-D electric field, to better understand how the electron foreshock beams are created.

3.2.3 Solar Wind Reconnection

The recent discovery of reconnection exhausts in the solar wind (Gosling et al. 2005) revealed a new laboratory for reconnection research. Extremely long X-lines, extending hundreds of Earth radii, have been reported in the solar wind (Phan et al. 2006, 2009; Gosling et al. 2007a). All extended X-line events reported so far were found in large magnetic shear current sheets where the reconnection exhausts were wide. However, there are many more low-magnetic shear (and much narrower) reconnection exhausts in the solar wind that can only be resolved by high time resolution plasma measurements (Gosling et al. 2007b). ARTEMIS, in conjunction with Wind, will allow the investigation of the extent of the reconnection X-lines for low magnetic shears, and establish whether their X-line extent scales with current sheet width as suggested in 3D simulations of reconnection (e.g. Shay et al. 2003). The multi-point, high-resolution measurements will also be used to investigate the structure of the reconnection exhaust as a function of the distance from the X-line.

3.2.4 Solar Wind Turbulence

Turbulence is a multi-scale phenomenon that mediates the transfer of energy, mass, and momentum. The presence of turbulence within the solar wind has been well established from studies of power spectra, probability distribution functions, scaling exponentials, Reynolds number determinations, etc. Most previous solar wind studies have not focused on the determination of the three fundamental turbulence sale lengths: the correlation scale, the Taylor scale, and the dissipation scale. The correlation scale is associated with the largest possible turbulent eddy scale size. The Taylor scale is the scale size within the inertial range at which viscous damping begins to become important for eddy damping, and the dissipation scale is the scale at which the turbulent eddies have been damped out and the particles are heated. Studies of these fundamental turbulence scales are important to help us understand cosmic ray modulation and particle scattering within the solar wind.

Studies of magnetic field fluctuations indicate that the correlation scale is on the order of several million kilometers and the Taylor scale on the order of a few thousand kilometers (Mattheaus et al. 1990, 2005; Dasso et al. 2005; Weygand et al. 2007, 2009a, 2009b). However, Mattheus et al. (1990), Dasso et al. (2005), and Weygand et al. (2009a, 2009b) have shown that solar wind turbulence is anisotropic and correlation scale can vary with the respect to the mean magnetic field direction. Mattheaus et al. (1990) attribute this feature to two different types of turbulence present within the solar wind: slab and quasi-two dimensional turbulence. Both Mattheus et al. (1990) and Dasso et al. (2005) used single spacecraft observation to show this anisotropy in the turbulent magnetic field fluctuations and determined the correlation scale by fitting an exponential function to autocorrelation values. Dasso et al. (2005) subdivided results by solar wind speed. They showed that the correlation scale is longest along the mean magnetic field direction in the slow solar wind ($<400~\mathrm{km\,s^{-1}}$) but longest in the perpendicular direction in the fast solar wind ($>500~\mathrm{km\,s^{-1}}$).

Weygand et al. (2009a, 2009b) employed spacecraft pairs within the solar wind to avoid the inherent assumption in previous studies that the solar wind magnetic field fluctuations are frozen into the flow. The Weygand et al. (2009b) study demonstrated systematic variations of the two-dimensional cross-correlation function for slow, intermediate, and fast solar wind speeds. However, two separate exponential fits were required to fit cross-correlation values across the full range of spacecraft separations observed. The first exponential fit the correlation values obtained for large ACE, Geotail, IMP-8, Wind, and Interball-1 interseparation distances, while the second exponential with a smaller decay length fit the much smaller interspacecraft separations for the Cluster spacecraft. Weygand et al. (2009a) hypothesized that this smaller scale resulted from instrumental differences or foreshock turbulence.

However, limited THEMIS solar wind magnetic field measurements indicate that the second exponential decay lengths at all solar wind speed ranges do not result from instrumental bias or foreshock turbulence, but rather represent real features in the solar wind. Figure 12 (Weygand et al. 2009b) displays the cross correlation values versus the spacecraft separation for various spacecraft pairs in the fast solar wind (>600 km/s). The blue values between the two vertical dashed lines are the cross-correlation values from the THEMIS B and C spacecraft.

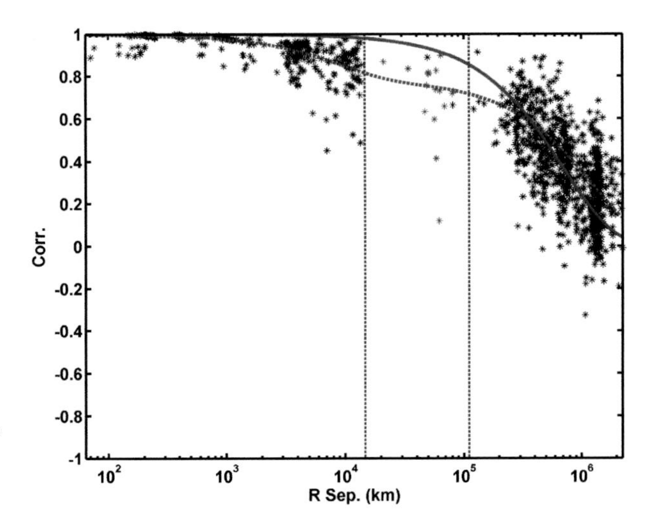

Fig. 12 Cross-correlation coefficients versus the distance separating spacecraft for the fast solar wind (>600 km/s) outside the foreshock. Cluster correlations cover the range of separations between 0 and $\sim$10,000 km in *black*. ACE, Wind, Geotail, IMP8, and Interball-1 correlations cover separations greater than 100,000 km. *Blue asterisks* indicate limited THEMIS B and C correlations between the two dashed vertical lines. The range curve results from a single exponential fit to all the data. The *dashed curve* represents the sum of two exponentials fit to all the data

While the THEMIS cross correlation values have been an enormous help, the physical meaning of the second exponential decay lengths remains unclear. It may be related to the hypothesis that turbulence is bound within solar wind flux tubes (Borovsky 2008). Large decay lengths are related to the walls of the solar wind flux tubes while the small decay lengths are associated with the correlation scale of the magnetic field turbulence within the flux tubes.

ARTEMIS will play a critical role in these investigations. Inter-separation distances at the Moon will range from 500 km to about 31,000 km, precisely filling in the poorly sampled region from 1.5×10^4 to 3.0×10^4 km where Fig. 12 indicates a transition between the two exponential fits.

3.2.5 Upstream Monitors

In lunar orbit, ARTEMIS will be an excellent monitor of the solar wind conditions upstream from the Earth's magnetosphere. Unlike Earth-orbiting satellites, ARTEMIS will remain outside the magnetosphere for extended periods of time, and will, for example, be able to observe the entire passage of a CME. As such, ARTEMIS will provide useful input to global simulations of the Earth's magnetosphere and services such as the Community Coordinated Modeling Center (CCMC). ARTEMIS will be closer to the Earth than existing L1 solar wind monitors, and therefore offers a more precise measurement of the exact solar wind input to the magnetosphere—such measurements will be of use to other teams studying the magnetosphere with both satellite- and ground-based experiments. Finally, by comparing ARTEMIS data to that from L1 monitors such as ACE and Wind, it will be possible to develop new models that advect L1 solar wind monitoring data to the Earth, thereby improving our understanding of how L1 monitoring data should be used to predict magnetospheric dynamics.

3.3 The Lunar Wake

As an essentially non-magnetic and non-conducting body with no ionosphere, the Moon absorbs most of the incident solar wind plasma, leaving a plasma void, or wake, within plasma flows (Schubert and Lichtenstein 1974). The plasma cavity formed downstream from the Moon represents one of the best natural plasma vacuums in the solar system, and an excellent basic physics laboratory for understanding the general process of plasma expansion into a vacuum, with applications ranging from low-altitude earth orbiting satellites to outer planet moons. Although the lunar wake has been studied since the Apollo era (Ness et al. 1967; Colburn et al. 1967), we still do not fully understand many aspects of its formation, dynamics, and refilling.

Nevertheless, our understanding of the wake has advanced with each new lunar mission. Far from being a passive region, the wake hosts a wide variety of dynamic phenomena (see Fig. 13). Wind, with a suite of modern plasma instruments, made a number of lunar flybys and discovered a wake extending to at least 25 R_L downstream (Clack et al. 2004), counter-streaming anisotropic ion beams refilling the wake along magnetic field lines (Ogilvie et al. 1996; Clack et al. 2004), and electrostatic and electromagnetic plasma waves in and around the wake (Kellogg et al. 1996; Farrell et al. 1996; Bale et al. 1997). Nozomi observed non-thermal ions and counter-streaming electrons upstream from the Moon (Futaana et al. 2001, 2003), possibly associated with wake processes. Geotail observed ULF waves associated with the lunar wake region (Nakagawa et al. 2003). Lunar Prospector discovered the important role of non-Maxwellian solar wind electrons in driving low-altitude wake refilling (Halekas et al. 2005). Kaguya, Chang'E, and Chandrayaan discovered ion scattering from the dayside surface (Saito et al. 2010; Wieser et al.

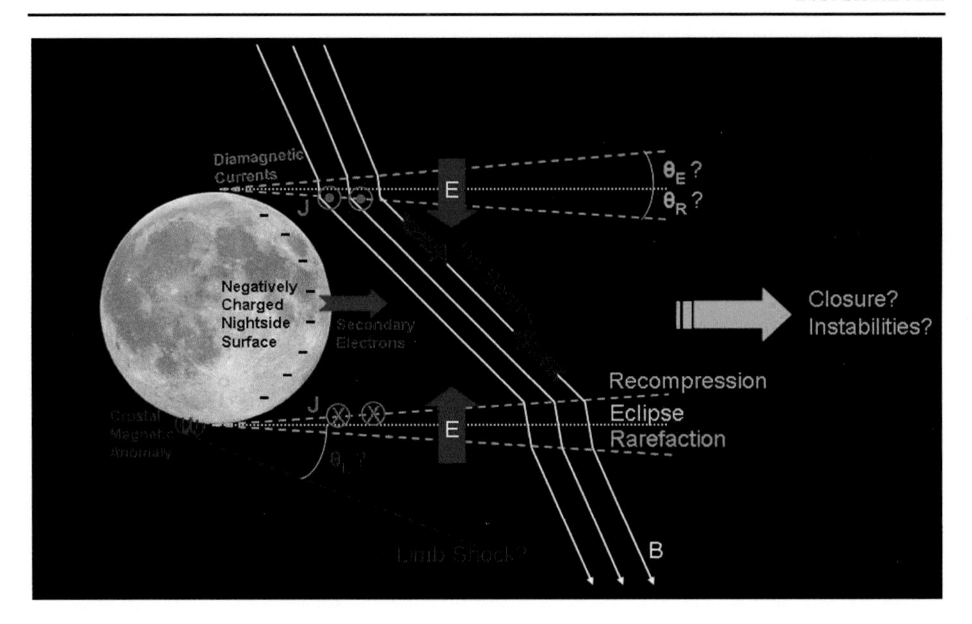

Fig. 13 Our current understanding of the electrodynamics of the Moon's interaction with the solar wind. The Moon carves out a cavity, known as the wake, in the oncoming solar wind. Ion beams stream inward to fill the cavity, which is bounded by rarefaction waves, diamagnetic currents, and inward-pointing electric fields. Crustal magnetic anomalies may launch shocks into the solar wind, while the nightside lunar surface charges negative and emits secondary electrons

2009), and observed gyrating protons (Type-I entry) and re-picked up scattered protons (Type-II entry) refilling the low altitude wake perpendicular to magnetic field lines (Nishino et al. 2009a, 2009b; Holmstrom et al. 2010; Wang et al. 2010).

While these observations have shed light on the global structure of the lunar wake and emphasized the need to treat it kinetically, our understanding of the physical nature of the wake and how it refills is limited primarily due to the relative paucity of in situ measurements made in the region. Early Explorer and Apollo satellites carried limited plasma instrumentation and the Lunar Prospector instrument package did not include ion detectors or electric field analyzers. Wind provided a relatively complete plasma data set from the lunar wake, but made only a handful of passes, leaving wake coverage far from complete. Nozomi and Geotail also have made only a very limited number of lunar flybys. Chandrayaan, Chang'E, and Kaguya only observed the wake over narrow ranges of distances from the Moon. To date, there has not been a dedicated mission to provide comprehensive coverage of the lunar wake over a wide range of distances and complete our understanding of its structure.

3.3.1 Structure of the Wake

Global hybrid simulations (kinetic ions, fluid electrons) confirm that kinetic effects are important. Figure 14, showing the formation of a plasma void in the lunar wake that is refilled by two beams counter-streaming along magnetic field lines in X–Y plane, demonstrates the complex structure of the lunar wake even for ambient magnetic field orientations that lie transverse to the wake axis. The same simulation also indicates that the lunar wake could extend well beyond 25 R_L downstream.

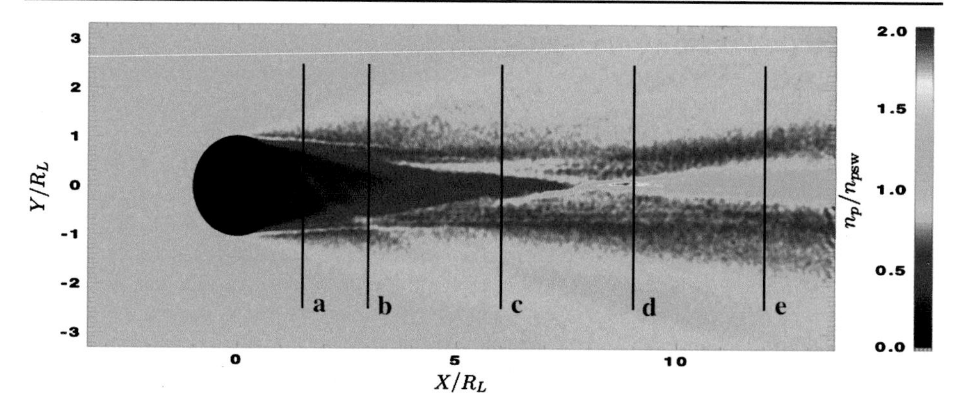

Fig. 14 Results from a global hybrid simulation for the interaction of the solar wind with the Moon. Proton densities, normalized to the value in the unperturbed solar wind, are shown in the X–Y plane containing the IMF, which points in the Y-direction. The results show the formation of a plasma void in the lunar wake that is refilled by two beams counter-streaming along magnetic field lines

As illustrated in Fig. 13, asymmetries in the lunar wake structure are expected whenever the magnetic field does not lie parallel to the wake, since diamagnetic current systems, and the resulting magnetic field perturbations, differ for perpendicular vs. parallel magnetic fields. In addition, limb compressions are often observed external to the wake cavity and rarefaction wave, located downstream from crustal magnetic anomalies (Russell and Lichtenstein 1975). Previous measurements suggest that these structures are compressional features (Siscoe et al. 1969) that propagate outward at magnetosonic wave velocities (Whang and Ness 1970). One expects them to propagate downstream from crustal magnetic field regions, producing a highly asymmetric structure external to the main wake cavity. However, no observations have clearly confirmed this supposition, or determined how far downstream these features propagate and to what degree they affect the structure of the wake interior to them. Many observations have found asymmetric wake characteristics on individual orbits (Ness et al. 1968), but single spacecraft cannot clearly distinguish between asymmetries and temporal variations.

The orbital coverage and complete plasma instrumentation of ARTEMIS will enable a comprehensive determination of the wake's extent and structure. The two-point measurements provided by ARTEMIS will allow unambiguous identification of asymmetries in the wake due to the perturbing influences of solar wind and crustal magnetic fields or other effects.

3.3.2 Wake Refilling and Dynamics

Previous missions have measured several modes of wake refilling, but we do not yet understand their relative importance or the interplay between them. Wind saw ion beams accelerated along magnetic field lines into the wake from the flanks (Ogilvie et al. 1996; Clack et al. 2004) several lunar radii downstream, implying a potential drop across the wake boundary that occurs as a natural consequence of the pressure gradient across the wake boundary and the difference in electron and ion thermal velocities. Simulations have contributed greatly to our understanding of this process, which refills the wake along magnetic field lines (Farrell et al. 1998; Birch and Chapman 2001, 2002; Kallio 2005; Travnicek et al. 2005; Kimura and Nakagawa 2008). Meanwhile, recent low altitude observations show that the wake also refills perpendicular to magnetic field lines (Nishino et al. 2009a, 2009b; Holmstrom et al. 2010; Wang et al. 2010). Only ARTEMIS, with its elliptical orbit, can unravel the interplay between parallel and perpendicular refilling processes, and determine their relative importance as a function of location in the wake.

The wake provides a rich laboratory for plasma waves and instabilities, with a broad spectrum of waves observed in the central wake (Kellogg et al. 1996), and even far upstream on magnetic field lines connected to the wake boundary (Nakagawa et al. 2003; Farrell et al. 1996; Bale et al. 1997). Waves observed to date include ion acoustic waves and Langmuir waves from instabilities related to differential ion/electron shadowing (Bale et al. 1997), and whistlers produced by beam instabilities near the wake boundary (Nakagawa et al. 2003; Farrell et al. 1996, 2008). Waves predicted but not conclusively observed before ARTEMIS include instabilities generated by counter-streaming electrons in the central wake (Birch and Chapman 2001, 2002), ion acoustic-like interactions generated by refilling ion beams in the central wake (Farrell et al. 1998), and low frequency electromagnetic turbulence with frequencies near the local proton gyrofrequency (Travnicek et al. 2005). Most recently, Kaguya has observed broad spectrum electrostatic turbulence associated with a two-stream instability formed by electrons pulled into the wake along field lines by the superabundance of charge from pickup ion Type II entry perpendicular to magnetic field lines (Nishino et al. 2010). We have barely explored this menagerie of plasma waves, and the instability growth mechanisms and wave-particle interactions remain far from understood. As illustrated in Fig. 15, the first lunar wake flyby by ARTEMIS revealed electrostatic oscillations associated with counter-streaming electrons in the central wake and the regions magnetically connected to it, providing the first confirmation for the predictions of Nakagawa et al. (2003) and Birch and Chapman (2001, 2002). See Halekas et al. (2011, this

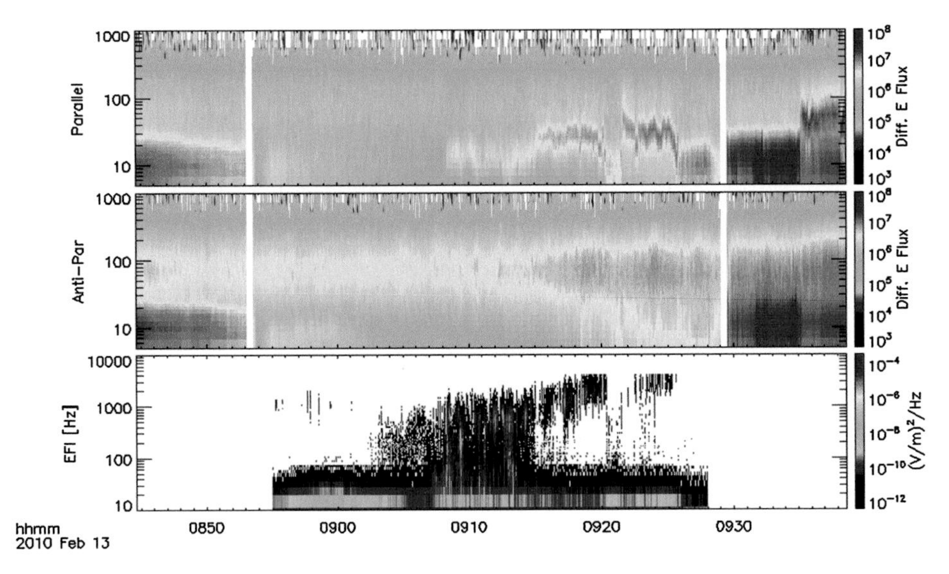

Fig. 15 Observations from the first lunar wake flyby by ARTEMIS probe P1 on Feb 13, 2010. The figure shows energy-time spectrograms for electrons streaming parallel (0–15° pitch angle) and anti-parallel (165–180°) to the magnetic field (in eV/(cm^2 s sr eV)), and an electric field wave frequency-time spectrogram. The anti-parallel streaming solar wind strahl population penetrates all the way through the wake from the exit side, while most of the core population is excluded from the central wake by the wake potential. A residual core population penetrates from the entry side of the wake, and is accelerated outward by the wake potential, forming a counter-streaming distribution that excites electrostatic oscillations

issue) for more details. Further passes will enable researchers to determine the distribution of wave activity within the wake.

3.3.3 Response to External Drivers

The wake structure will certainly vary in dynamic response to external drivers. Several studies (Ogilvie and Ness 1969; Colburn et al. 1971; Whang and Ness 1972; Halekas et al. 2005) provide tantalizing hints how the wake responds to changing solar wind conditions, but the limited data and orbital coverage of the previous missions constrains our knowledge of this response to external drivers. Hybrid simulations have shown that the orientation of the IMF with respect to the solar wind flow direction affects both the downstream extent and symmetry of the lunar wake (Travnicek et al. 2005).

Although the Moon spends the majority of its time in the solar wind, it also spends ~5 days each month in the terrestrial magnetosphere, where the plasma environment differs greatly from that in the solar wind. Magnetic field configurations and flow speeds in the plasma sheet are highly variable, which should result in lunar wakes very different from those in the solar wind. For example, reconnection beyond 60 R_E should result in sunward plasma sheet plasma flows and a sunward-orientated wake.

Using dual-probe measurements, ARTEMIS will determine the response of the lunar wake to solar wind drivers. ARTEMIS will provide the first detailed measurement of the lunar wake plasma environment within the terrestrial magnetosphere.

4 Concluding Remarks

From vantage points near and in orbit about the Moon, ARTEMIS will provide the comprehensive two-point particle and electromagnetic field observations needed to chart and understand a host of planetary phenomena. Observations of pick-up ions will be used to determine the composition of the exosphere and hence the lunar surface as a function of external conditions. Electric fields that can loft dust in the vicinity of the Moon will be measured both directly and remotely. The structure of crustal magnetic anomalies that shield some regions of the Moon while creating mini-magnetospheres complete with bow shocks in the oncoming solar wind will be mapped.

The same lunar vantage points offer an opportunity to address a series of longstanding heliospheric research problems. ARTEMIS will map the structure of the Earth's magnetotail, determining its structure and the characteristics of magnetic reconnection, particle energization, and turbulence as function of solar wind conditions. While in the solar wind, the two spacecraft will provide the observations needed to understand reconnection at interplanetary discontinuities and particle acceleration at both the Earth's bow shock and interplanetary shocks. They will also serve as excellent monitors of the solar wind input into the magnetosphere for other forthcoming NASA and international missions. Finally, the ARTEMIS spacecraft will provide the comprehensive observations needed to understand the steady-state and time-dependent structure of and processes within the lunar wake.

The novel repurposing of the outermost two THEMIS spacecraft demonstrates the creativity of the THEMIS/ARTEMIS team at its best. It has set the stage for potentially dramatic advances in our understanding of the Moon and its environment. This will be a joint undertaking of both the THEMIS/ARTEMIS team and the international community, for all of the observations returned by ARTEMIS will become immediately available via the mission's web site, themis.ssl.Berkeley.edu, just as is the case for THEMIS.

Acknowledgements Work at UCB and UCLA was supported by NASA Contract NAS5-02099. Work at NASA/GSFC was supported by THEMIS MO&DA. J.P.E. is supported by an STFC Advanced Fellowship at Imperial College.

References

K.A. Anderson et al., Space Sci. Instrum. **1**, 439 (1975)
V. Angelopoulos et al., J. Geophys. Res. **97**, 4027 (1992). doi:10.1029/91JA02701
V. Angelopoulos et al., J. Geomagn. Geoelectr. **48**, 629 (1996)
V. Angelopoulos, Space Sci. Rev. **141** (2008). doi:10.1007/s11214-008-9336-1
V. Angelopoulos, Space Sci. Rev. (2011) (this issue). doi:10.1007/s11214-010-9687-2
W.I. Axford, C.O. Hines, Can. J. Phys. **39**, 1433 (1961)
I.A. Axford et al., in *Proceedings of the 15th International Cosmic Ray Conf.*, vol. 11 (1977), p. 132
S.D. Bale et al., Geophys. Res. Lett. **24**, 1427 (1997)
S.D. Bale et al., Geophys. Res. Lett. **26**, 1573 (1999)
W. Baumjohann et al., J. Geophys. Res. **94**, 6597 (1989). doi:10.1029/JA094iA06p06597
W. Baumjohann et al., J. Geophys. Res. **95**, 3801 (1990). doi:10.1029/JA095iA04p03801
A.R. Bell, Mon. Not. R. Astron. Soc. **182**, 147 (1978a)
A.R. Bell, Mon. Not. R. Astron. Soc. **182**, 443 (1978b)
P.C. Birch, S.C. Chapman, Phys. Plasmas **8**, 4551 (2001)
P.C. Birch, S.C. Chapman, Phys. Plasmas **9**, 1785 (2002)
R.D. Blandford, J.P. Ostriker, Astrophys. J. **221**, L29 (1978)
J.E. Borovsky, J. Geophys. Res. **113** (2008). doi:10.1029/2007JA012684
T.K. Breus et al., Adv. Space Res. **36**, 2043 (2005)
D. Burgess et al., Space Sci. Rev. **118**, 205 (2005)
D. Burgess, Lect. Notes Phys. **725**, 161 (2007)
C.C. Chaston et al., Phys. Rev. Lett. **99**, 175004 (2007)
C.C. Chaston et al., Geophys. Res. Lett. **35** (2008). doi:10.1029/2008GL033601
S.-H. Chen, M.G. Kivelson, Geophys. Res. Lett. **20**, 2699 (1993)
S.M. Cisowski et al., J. Geophys. Res. Suppl. **88**, A691 (1983)
D. Clack et al., Geophys. Res. Lett. **31** (2004). doi:10.1029/2003GL018298
J.B. Cladis et al., J. Geophys. Res. **99**, 53 (1994)
D.S. Colburn et al., Science **158**, 1040 (1967)
D.S. Colburn et al., J. Geophys. Res. **76**, 2940 (1971)
P.J. Coleman et al., Proc. Lunar Sci. Conf. **3**, 2271 (1972)
J.E.P. Connerney et al., Science **284**, 794 (1999)
J.E.P. Connerney et al., Space Sci. Rev. **1**, 1 (2004)
D.R. Criswell, in *Photon and Particle Interaction in Space*, ed. by R.J.L. Grard (Reidel, Dordrecht, 1973), p. 545
S. Dasso et al., Astrophys. J. **635**, L181 (2005)
G. Delory et al., in *40th LPSC*, id. 2025, 2009
M. Delva et al., Geophys. Res. Lett. **35** (2008). doi:10.1029/2007GL032594
M.I. Desai et al., J. Geophys. Res. **105**, 61 (2000)
M. Dougherty et al., Science **311**, 1406 (2006)
J.F. Drake et al., Nature **443**, 553 (2006). doi:10.1038/nature05116
J.F. Drake et al., J. Geophys. Res. **114** (2009). doi:10.1029/2008JA013701
J.W. Dungey, Phys. Rev. Lett. **6**, 47 (1961)
Dyal, et al., Rev. Geophys. Space Phys. **12**, 568 (1974)
J.P. Eastwood et al., Space Sci. Rev. **118**, 41 (2005)
D.H. Fairfield et al., J. Geophys. Res. **105**, 21159 (2000)
W.M. Farrell et al., Geophys. Res. Lett. **23**, 1271 (1996)
W.M. Farrell et al., J. Geophys. Res. **103**, 23653 (1998)
W.M. Farrell et al., Geophys. Res. Lett. **34** (2007). doi:10.1029/2007GL029312
W.M. Farrell et al., Geophys. Res. Lett. **35** (2008). doi:10.1029/2007GL032653
M.H. Farris, C.T. Russell, J. Geophys. Res. **99**, 17681 (1994)
J.A. Fedder, J.G. Lyon, J. Geophys. Res. **100**, 3623 (1995)
W.C. Feldman et al., J. Geophys. Res. **90**, 233 (1985)
B. Flynn, M. Mendillo, Science **261**, 184 (1993)
H.U. Frey et al., Nature **426**, 533 (2003)

M. Fujimoto, T. Terasawa, J. Geophys. Res. **99**, 8601 (1994)
M. Fuller, Rev. Geophys. Space Phys. **12**, 23 (1974)
M. Fuller, S. Cisowski, in *Geomagnetism*, vol. 2, ed. by J. Jacobs (Academic Press, London, 1987), p. 307
Y. Futaana et al., J. Geophys. Res. **106**, 18729 (2001)
Y. Futaana et al., J. Geophys. Res. **108**, 1025 (2003). doi:10.1029/2002JA009366
V.L. Galinsky, B.U.Ö. Sonnerup, Geophys. Res. Lett. **21**, 2247 (1994)
J. Giacalone, Astrophys. J. **609**, 452 (2004)
G.R. Gladstone et al., LPSC, abstract 2277 (2010)
T.I. Gombosi et al., in *Physics of Space Plasmas*, ed. by T. Chang, J.R. Jasperse (MIT, Cambridge, 1998), p. 121
T.I. Gombosi et al., J. Geophys. Res. **105**, 13141 (2000)
J.T. Gosling et al., J. Geophys. Res. **94**, 3555 (1989)
J.T. Gosling et al., J. Geophys. Res. **110** (2005). doi:10.1029/2004JA010809
J.T. Gosling et al., Geophys. Res. Lett. **34** (2007a). doi:10.1029/2006GL029033
J.T. Gosling et al., Geophys. Res. Lett. **34** (2007b). doi:10.1029/2007GL030706
F.S. Grant, G.F. West, *Interpretation Theory in Applied Geophysics* (McGraw-Hill, New York, 1965)
J.M. Grebowsky et al., Adv. Space Res. **33**, 176 (2004)
R.E. Grimm, NRC planetary science white paper. NRC, Washington, DC, http://www.psi.edu/decadal/ 2009
R.E. Grimm, G.T. Delory, NLSI Lunar Science Conf., abstract #2075 (2008)
R.E. Grimm, H.Y. McSween, 40th LPSC XL, abstract #1958 (2009)
P.N. Guzdar et al., J. Geophys. Res. **106**, 275 (2001)
J.S. Halekas, PhD thesis, UC Berkeley, 2003
J.S. Halekas et al., J. Geophys. Res. **106**, 27841 (2001)
J.S. Halekas et al., Geophys. Res. Lett. **29** (2002a). doi:10.1029/2001GL014428
J. Halekas et al., Geophys. Res. Lett. **29** (2002b). doi:10.1029/2001GL013924
J.S. Halekas et al., J. Geophys. Res. **110** (2005). doi:10.1029/2004JA010991
J.S. Halekas et al., Geophys. Res. Lett. **33** (2006a). doi:10.1029/2006GL025931
J.S. Halekas et al., Geophys. Res. Lett. **33** (2006b). doi:10.1029/2006GL027684
J.S. Halekas et al., Geophys. Res. Lett. **34** (2007). doi:10.1029/2006GL028517
J.S. Halekas et al., Planet. Space Sci. **56**, 941 (2008a). doi:10.1016/j.pss.2008.01.008
J.S. Halekas et al., J. Adv. Space Res. **41**, 1319 (2008b). doi:10.1016/j.asr.2007.04.003
J.S. Halekas et al., J. Geophys. Res. **113** (2008c). doi:10.1029/2008JA013194
J.S. Halekas et al., Space Sci. Rev. (2011) (this issue). doi:10.1007/s11214-010-9738-8
M. Hapgood, Ann. Geophys. **25**, 2037 (2007). doi:10.5194/angeo-25-2037-2007
E.M. Harnett, R. Winglee, J. Geophys. Res. **105**, 24997 (2000)
E.M. Harnett, R. Winglee, J. Geophys. Res. **108** (2003). doi:10.1029/2002JA009617
E.C. Hartle, R. Killen, Geophys. Res. Lett. **33** (2006). doi:10.1029/2005GL024520
E.C. Hartle, E.C. Sittler Jr., J. Geophys. Res. **112** (2007). doi:10.1029/2006JA012157
E.C. Hartle, G.E. Thomas, J. Geophys. Res. **79**, 1519 (1974)
H. Hasegawa et al., Ann. Geophys. **22**, 1251 (2004)
H. Hasegawa et al., Adv. Space Res. **36**, 1772 (2005)
H. Hasegawa et al., J. Geophys. Res. **111** (2006). doi:10.1029/2006JA011728
H. Hasegawa et al., Geophys. Res. Lett. **35** (2008). doi:10.1029/2008GL034767
M. Hilchenbach et al., in *Proceedings of Solar Wind Seven*, ed. by E. Marsch, G. Schwenn (Pergamon Press, Oxford, 1991)
M. Hilchenbach et al., Adv. Space Res. **13**, 321 (1993)
M.M. Holmström et al., J. Geophys. Res., in press (2010). doi:10.1029/2009JA014843
L.L. Hood, J. Geophys. Res. **97**, 18275 (1992)
L.L. Hood, N.A. Artemieva, Icarus **193**, 485 (2008)
L.L. Hood, G. Schubert, Science **204**, 49 (1980)
L.L. Hood, C.P. Sonett, Geophys. Res. Lett. **9**, 37 (1982)
L.L. Hood, Z. Huang, J. Geophys. Res. **96**, 9837 (1991)
L.L. Hood, R.W. Williams, in *LPSC 19* (1989), p. 19
L.L. Hood et al., J. Geophys. Res. **86**, 1055 (1981)
L.L. Hood et al., J. Geophys. Res. **87**, 5311 (1982)
L.L. Hood et al., Geophys. Res. Lett. **26**, 2327 (1999)
L.L. Hood et al., J. Geophys. Res. **106**, 27825 (2001)
M. Hoshino et al., J. Geophys. Res. **106**, 25979 (2001). doi:10.1029/2001JA900052
A. Ieda et al., J. Geophys. Res. **103**, 4453 (1998)
ILN, Final Report of the Science Definition Team for the ILN Anchor Nodes. NASA, Washington, DC, 2009, http://lunarscience.arc.nasa.gov/pdf/ILN_Final_Report.pdf

E. Kallio, Geophys. Res. Lett. **32** (2005). doi:10.1029/2004GL021989
P.J. Kellogg et al., Geophys. Res. Lett. **23**, 1267 (1996)
C.F. Kennel et al., J. Geophys. Res. **91**, 11917 (1986)
K.K. Khurana et al., Nature **395**, 777 (1998)
S. Kimura, T. Nakagawa, Earth Planets Space **60**, 591 (2008)
E. Kirsch et al., Adv. Space Res. **20**, 845 (1997)
A. Kis et al., Geophys. Res. Lett. **31** (2004). doi:10.1029/2004GL020759
M.G. Kivelson et al., Adv. Space Res. **16**, 59 (1995)
M.G. Kivelson et al., Science **274**, 396 (1996)
E.A. Kronberg et al., J. Geophys. Res. **114** (2009). doi:10.1029/2008JA013754
S.H. Lai, L.H. Lyu, J. Geophys. Res. **111** (2006). doi:10.1029/2004JA010724
M.A. Lee, J. Geophys. Res. **88**, 6109 (1983)
W. Li et al., J. Geophys. Res. **113** (2008). doi:10.1029/2007JA012604
R.P. Lin et al., J. Geophys. Res. **79**, 489 (1974)
R.P. Lin et al., Proc. Lunar Sci. Conf. **7**, 2691 (1976)
R.P. Lin et al., Icarus **74**, 529 (1988)
R.P. Lin et al., Science **281**, 1480 (1998)
J.E. McCoy, Proc. Lunar Sci. Conf. **7**, 1087 (1976)
J.E. McCoy, Criswell, Proc. Lunar Sci. Conf. **5**, 2991 (1974)
U. Mall et al., Geophys. Res. Lett. **25**, 3799 (1998)
R.H. Manka, in *Photon and Particle Interactions with Surfaces in Space*, ed. by R.J.L. Grard (Reidel, Dordrecht, 1973), p. 347
W.H. Matthaeus et al., J. Geophys. Res. **95**, 20673 (1990)
W.H. Matthaeus et al., Phys. Rev. Lett. **95**, 231101 (2005)
Y. Matsumoto, K. Seki, J. Geophys. Res. **112** (2007). doi:10.1029/2006JA012114
M. Mendillo et al., Geophys. Res. Lett. **18**, 2907 (1991)
M. Mendillo et al., Icarus **137**, 13 (1999)
D.L. Mitchell et al., Icarus **194**, 401 (2008). doi:10.1016/j.icarus.2007.10.027
A. Miura, J. Geophys. Res. **89**, 801 (1984)
A. Miura, J. Geophys. Res. **97**, 10655 (1992)
A. Miura, Phys. Plasmas **4**, 2871 (1997)
E. Möbius et al., Geophys. Res. Lett. **13**, 1372 (1986)
T. Nakagawa et al., Earth Planets Space **55**, 569 (2003)
T.K.M. Nakamura et al., Phys. Plasmas **17**, 042119 (2010)
N.F. Ness et al., J. Geophys. Res. **72**, 5769 (1967)
N.F. Ness et al., J. Geophys. Res. **73**, 3421 (1968)
P.T. Newell et al., J. Geophys. Res. **102**, 127 (1997)
A. Nishida et al., Geophys. Res. Lett. **22**, 2453 (1995)
M.N. Nishino et al., Geophys. Res. Lett. **36** (2009a). doi:10.1029/2009GL039049
M.N. Nishino et al., Geophys. Res. Lett. **36** (2009b). doi:10.1029/2009GL039444
M.N. Nishino et al., Geophys. Res. Lett. **37** (2010). doi:10.1029/2010GL043948
M.N. Nishino et al., Planet. Space Sci. **59** (2011). doi:10.1016/j.pss.2010.03.011
T. Nitter et al., J. Geophys. Res. **103**, 6605 (1998)
K.W. Ogilvie, N.F. Ness, J. Geophys. Res. **74**, 4123 (1969)
K.W. Ogilvie et al., Geophys. Res. Lett. **10**, 1255 (1996)
T. Ogino et al., IEEE Trans. Plasma Sci. **20**, 817 (1992)
M. Øieroset et al., J. Geophys. Res. **105**, 25247 (2000)
M. Øieroset et al., Phys. Rev. Lett. **89** (2002). doi:10.1103/PhysRevLett.89.195001
M. Øieroset et al., Geophys. Res. Lett. **31** (2004). doi:10.1029/2004GL020321
M. Øieroset et al., J. Geophys. Res. **113** (2008). doi:10.1029/2007JA012679
N. Omidi et al., J. Geophys. Res. **107**, 1487 (2002). doi:10.1029/2002JA009441
W.R. Paterson et al., J. Geophys. Res. **104**, 22779 (1999)
T.-D. Phan et al., Nature **439**, 175 (2006)
T.-D. Phan et al., Geophys. Res. Lett. **36** (2009). doi:10.1029/2009GL037713
A. Poppe, M. Horányi, J. Geophys. Res. **115** (2010). doi:10.1029/2010JA015286
A.E. Potter, T.H. Morgan, Science **241**, 675 (1988)
M. Pulupa, S.D. Bale, Astrophys. J. **676**, 1330 (2008)
J. Raeder, J. Geophys. Res. **104**, 17357 (1999)
J. Raeder, J. Geophys. Res. **105**, 13149 (2000)
J. Raeder, Ann. Geophys. **24**, 381 (2006)
J. Raeder et al., Geophys. Res. Lett. **22**, 349 (1995)

J.J. Rennilson, D.R. Criswell, Moon **10**, 121 (1974)
N.C. Richmond et al., J. Geophys. Res. **110** (2005). doi:10.1029/2005JA002405
C.T. Russell, R.C. Elphic, Space Sci. Rev. **22**, 681 (1978)
C.T. Russell, X. Blanco-Cano, J. Atmos. Sol.-Terr. Phys. **69**, 1723 (2007)
C.T. Russell, B.R. Lichtenstein, J. Geophys. Res. **80**, 4700 (1975)
C.T. Russell et al., in *LPSC 12th* (Pergamon, New York, 1982), p. 831
C.T. Russell et al., Geophys. Res. Lett. **17**, 897 (1990)
A.E. Saal et al., Nature **454**, 192 (2008)
Y. Saito et al., Geophys. Res. Lett. **35** (2008). doi:10.1029/2008GL036077
Y. Saito et al., Space Sci. Rev. **154** (2010). doi:10.1007/s11214-010-9647-x
E.T. Sarris et al., J. Geophys. Res. **92**, 12083 (1987)
G. Schubert, B.R. Lichtenstein, Rev. Geophys. Space Phys. **12**, 592 (1974)
G. Schubert, K. Schwartz, J. Geophys. Res. **77**, 76 (1972)
A. Schuster, H. Lamb, Philos. Trans. R. Soc. A **180**, 467 (1889)
M.A. Shay et al., Geophys. Res. Lett. **30**, 1345 (2003). doi:10.1029/2002GL016267
D.G. Sibeck, G.L. Siscoe, J. Geophys. Res. **89**, 10709 (1984)
D.G. Sibeck et al., J. Geophys. Res. **90**, 4011 (1985)
D.G. Sibeck et al., in *Magnetotail Physics*, ed. by A.T.Y. Lui (Johns Hopkins Press, Baltimore, 1987), p. 7
F. Simpson, K. Bahr, *Practical Magnetotellurics* (Cambridge University Press, Cambridge, 2005)
G.L. Siscoe et al., J. Geophys. Res. **74**, 59 (1969)
J.A. Slavin et al., Geophys. Res. Lett. **26**, 2897 (1999)
C.P. Sonett et al., Proc. Lunar Sci. Conf 3rd **2**, 2309 (1972)
C.P. Sonett, Rev. Geophys. Space Phys. **20**, 411 (1982)
D.J. Southwood, Planet. Space Sci. **16**, 587 (1968)
S.A. Stern, Rev. Geophys. **37**, 453 (1999)
T.J. Stubbs et al., Adv. Space Res. **37**, 59 (2006)
T.J. Stubbs et al., in *Dust in Planetary Systems*, vol. SP-643, ed. by H. Krüger, A.L. Graps (ESA, Noordwijk, 2007), p. 181
K. Takagi et al., J. Geophys. Res. **111** (2006). doi:10.1029/2006JA011631
T. Tanaka et al., Geophys. Res. Lett. **22** (2009). doi:10.1029/2009GL040682
T. Terasawa, Adv. Space Res. **15**, 53 (1995)
T. Terasawa, Prog. Theor. Phys. Suppl. **151**, 95 (2003)
K.J. Trattner et al., J. Geophys. Res. **99**, 13389 (1994)
P. Travnicek et al., Geophys. Res. Lett. **32** (2005). doi:10.1029/2004GL022243
A.L. Tyler et al., Geophys. Res. Lett. **15**, 1141 (1988)
A. Usadi et al., J. Geophys. Res. **98**, 7503 (1993)
P. Vernazza et al., Astron. Astrophys. **451**, L43 (2006)
F.J. Vine, D.H. Matthews, Nature **199**, 947 (1963)
F.J. Vine, J.T. Wilson, Science **150**, 485 (1965)
K. Vozoff, in *Electromag. Meth. Appl. Geophys*, vol. 2, ed. by M.N. Nabighian (Soc. Explor Geophys., Tulsa, 1991), Part B, p. 641
K. Watanabe, T. Sato, J. Geophys. Res. **95**, 75 (1990)
X.-D. Wang et al., Geophys. Res. Lett. **37** (2010). doi:10.1029/2010GL042891
J.M. Weygand et al., J. Geophys. Res. **112** (2007). doi:10.1029/2007JA012486
J.M. Weygand et al., J. Geophys. Res. **114** (2009a). doi:10.1029/2008JA013766
J.M. Weygand et al., EOS Trans. **90** (2009b)
Y.C. Whang, N.F. Ness, J. Geophys. Res. **75**, 6002 (1970)
Y.C. Whang, N.F. Ness, J. Geophys. Res. **77**, 1109 (1972)
M.A. Wieczorek, B.P. Weiss, Lunar Planet. Inst. Sci. Conf. Abstr. **41**, 1625 (2010)
M.A. Wieczorek et al., Rev. Mineral. Geochem. **60**, 221 (2006)
P. Wiedelt, J. Geophys. **38**, 257 (1972)
M. Wieser et al., Planet. Space Sci. (2009). doi:10.1016/j.pss.2009.09.012
M. Wieser et al., Geophys. Res. Lett. **37** (2010). doi:10.1029/2009GL041721
J.K. Wilson et al., Geophys. Res. Lett. **30** (2003). doi:10.1029/2003GL017443
J.K. Wilson et al., J. Geophys. Res. **111** (2006). doi:10.1029/2005JA011364
P. Wurtz et al., Icarus **191**, 486 (2007)
S. Yokota, Y. Saito, Earth Planets Space **57**, 281 (2005)
S. Yokota et al., Geophys. Res. Lett. **36** (2009). doi:10.1029/2009GL038185

DOI 10.1007/978-1-4614-9554-3_4
Reprinted from *Space Science Reviews* Journal, DOI 10.1007/s11214-012-9869-1

ARTEMIS Mission Design

Theodore H. Sweetser · Stephen B. Broschart · Vassilis Angelopoulos · Gregory J. Whiffen · David C. Folta · Min-Kun Chung · Sara J. Hatch · Mark A. Woodard

Received: 19 May 2011 / Accepted: 17 February 2012 / Published online: 24 March 2012

Abstract The ARTEMIS mission takes two of the five THEMIS spacecraft beyond their prime mission objectives and reuses them to study the Moon and the lunar space environment. Although the spacecraft and fuel resources were tailored to space observations from Earth orbit, sufficient fuel margins, spacecraft capability, and operational flexibility were present that with a circuitous, ballistic, constrained-thrust trajectory, new scientific information could be gleaned from the instruments near the Moon and in lunar orbit. We discuss the challenges of ARTEMIS trajectory design and describe its current implementation to address both heliophysics and planetary science objectives. In particular, we explain the challenges imposed by the constraints of the orbiting hardware and describe the trajectory solutions found in prolonged ballistic flight paths that include multiple lunar approaches, lunar flybys, low-energy trajectory segments, lunar Lissajous orbits, and low-lunar-periapse orbits. We conclude with a discussion of the risks that we took to enable the development and implementation of ARTEMIS.

Keywords ARTEMIS · THEMIS · Low-energy transfer · Lissajous orbits · Lunar science · Lunar mission · Heliophysics · Magnetosphere

1 Introduction

Time History of Events and Macroscale Interactions during Substorms (THEMIS) is a very successful NASA Explorer mission launched in February of 2007 to advance our understanding of magnetic substorms, a space weather phenomenon in the Earth's magnetosphere (Angelopoulos 2008). The mission consists of five identical Earth-orbiting spacecraft

T.H. Sweetser (✉) · S.B. Broschart · V. Angelopoulos · G.J. Whiffen · M.-K. Chung · S.J. Hatch
Jet Propulsion Laboratory, California Institute of Technology, 4800 Oak Grove Dr., M/S, 301-121, Pasadena, CA 91109, USA
e-mail: Ted.Sweetser@jpl.nasa.gov

D.C. Folta · M.A. Woodard
Goddard Space Flight Center, Greenbelt, MD, USA

(probes) equipped with particle and field instruments (Harvey et al. 2008). As of the time of this writing, the baseline mission science objectives have been achieved, and all five probes (and their instruments) are fully functional.

In February 2008 ARTEMIS, the Acceleration, Reconnection, Turbulence and Electrodynamics of the Moon's Interaction with the Sun mission, was proposed to the NASA Heliophysics Senior Review (Angelopoulos and Sibeck 2008) as an extension to the THEMIS mission. It was approved for development in May of that year. The ARTEMIS mission proposed to send the two outermost THEMIS probes, P1 and P2 (also referred to as THEMIS-B and THEMIS-C), to lunar orbits by way of two circuitous transfers that take about one and a half years each. The goals of the mission as proposed in 2008 were to use the Moon as an anchor for the ARTEMIS probes to conduct studies of Earth's magnetotail and solar wind from approximately 60 Earth radii and to study the lunar wake and its refilling as a function of the upstream solar wind. ARTEMIS two-point measurements open a new vantage point to phenomena previously studied by single-spacecraft missions. In particular, when solar wind measurements are made simultaneously by one probe in the lunar wake and the second from various locations just upstream of the lunar wake, accurate comparisons of wake phenomena with upstream variations can be made.

The ARTEMIS proposal represented the combined efforts of the THEMIS science team led by the PI at UCLA, the THEMIS Mission Operations team led by the Mission Operations Manager at the University of California Berkeley's Space Science Laboratory (UCB-SSL), the NASA Goddard Space Flight Center (GSFC), and the Jet Propulsion Laboratory at the California Institute of Technology (JPL). Two earlier reports (Broschart et al. 2009; Woodard et al. 2009) describe the preliminary mission design as proposed in 2008; portions of this paper are taken from those reports. This paper presents the evolution of the trajectory design to the trajectory being flown today, only a few months after lunar orbit delivery.

Numerous challenges were inherent to the ARTEMIS mission's trajectory design because of the constrained capabilities of the THEMIS probes. Limited fuel remained after the THEMIS baseline mission was completed. Thruster configuration limits thrust directions to one hemisphere. Additionally, an on-off thruster duty cycle imposed due to the spinning of the probe bus restricts effective thrust to less than a newton in the spin plane, i.e., for maneuver directions near the ecliptic plane. Maneuvers cannot be done in shadow because accurate pulse timing relies on sun-sensor data. Telecommunications with the probes were limited to a range of about two million kilometers. Finally, the probes can only withstand up to a 4-hour shadow. Had nothing been done at the end of the THEMIS baseline mission, long eclipses (>8 hr) would have neutralized P1 by March 2010 (Angelopoulos 2010). This became a very significant driver for proposing the ARTEMIS mission.

In Sect. 2 we describe the capabilities and orbit configuration of the THEMIS probes at the end of their baseline mission. In Sect. 3 we outline the history of the ARTEMIS mission design concept as it followed the mission's programmatic evolution. Section 4 outlines the science goals and orbit design goals of the mission. The remainder of the paper describes the design of the trajectories that have taken P1 and P2 from eccentric, high-altitude Earth orbits into lunar orbits that satisfy the science objectives. Figure 1 shows the ARTEMIS trajectory design used to send P1 and P2 from their respective Earth orbits at the start of ARTEMIS maneuvers into lunar Lissajous orbits. Section 5 presents the most up-to-date ARTEMIS mission design. Section 6 describes the current mission status, including ongoing trade studies. Section 7 is a retrospective on the challenges and enabling attributes of the mission design effort.

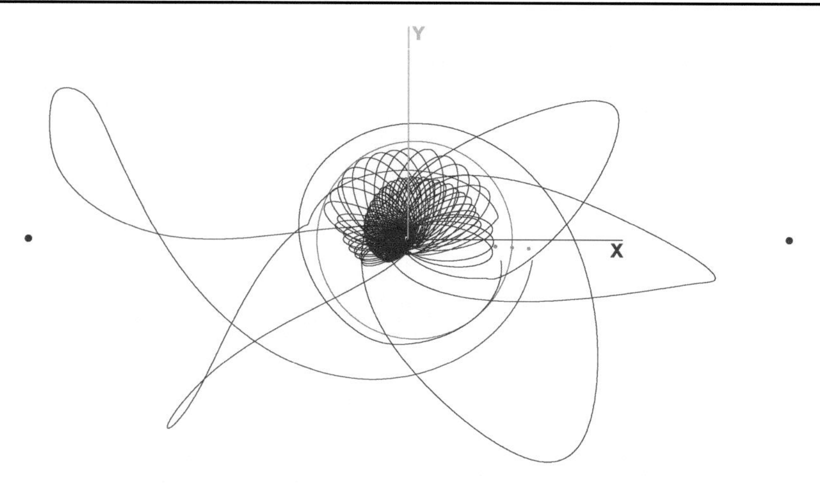

Fig. 1 ARTEMIS trans-lunar trajectories in the ecliptic plane. The coordinate frame here rotates such that the Sun is always to the left. The *red line* shows the P1 trajectory; the *blue line* shows the P2 trajectory. The Earth is at the center of the figure, and the Moon's orbit is shown in gray. The *blue dots* are the Sun-Earth L1 and L2 Lagrange points; the *gray dots* are the Moon and the Earth-Moon L1 and L2 points at a particular epoch

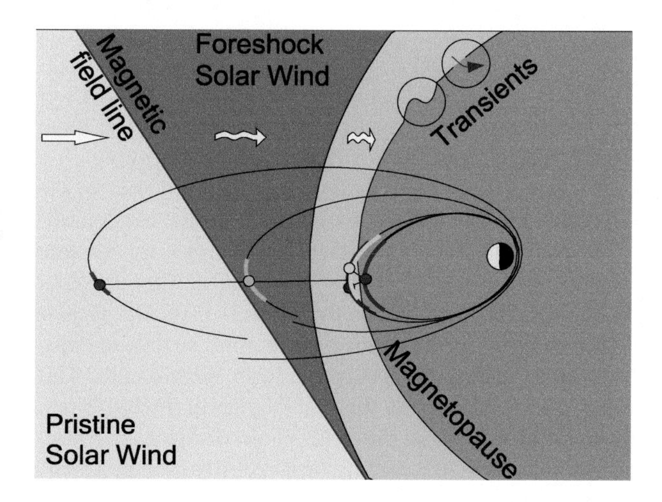

Fig. 2 THEMIS mission orbit configuration. *Filled circles* represent THEMIS probe locations during a dayside conjunction (*red*: P1 4-day orbit, *green*: P2 2-day orbit, *black*: P3 1-day orbit, *blue*: P4 1-day orbit, *pink*: P5 1-day orbit). The orbit geometries are indicated by *black lines*

2 Spacecraft Overview

On February 17th, 2007, the five THEMIS probes were launched on a Delta-II 7925 rocket into a 1.3-day Earth orbit with perigee at 437 km altitude and apogee at ~87500 km altitude (Angelopoulos 2008). Based on initial on-orbit data—in particular, better link margin performance—THEMIS-B was assigned to a 4-day orbit and designated "P1", and THEMIS-C was assigned to a 2-day orbit and designated "P2". THEMIS-D, E, and A were assigned to 1-day orbits, becoming P3, 4 and 5, respectively, per the mission design plan (Frey et al. 2008) required to achieve THEMIS mission science goals (Fig. 2) (Angelopoulos 2008). After 29 months in orbit, the two outermost probes, P1 and P2, were called on to journey to the Moon as part of the ARTEMIS mission.

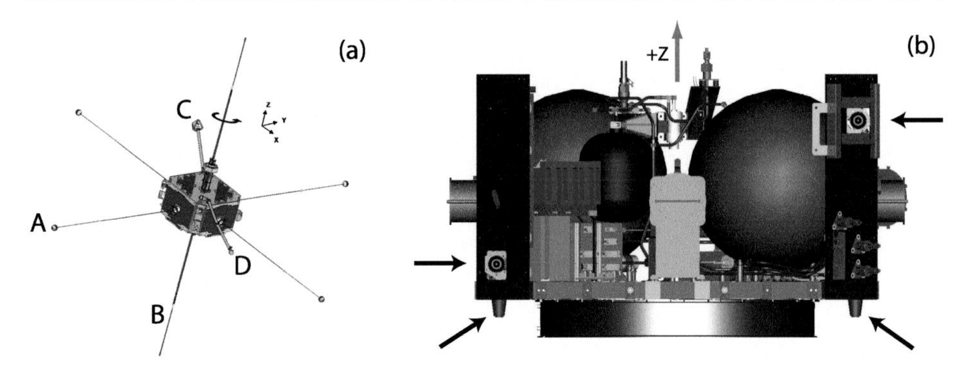

Fig. 3 THEMIS/ARTEMIS probe configuration. The probe buses were manufactured by ATK Space Systems (formerly Swales Aerospace), and the instruments were manufactured under the leadership of the University of California, Berkeley with both US and international collaborators. (**a**) On-orbit configuration with booms deployed, adapted from Auslander et al. (2008): *A*—four 20 m long radial EFI booms; *B*—two 5 m long axial EFI booms; *C*—1 m long SCM boom; *D*—2 m long FGM boom (http://www.nasa.gov/images/content/164405main_THEMIS-Spacecraft_bus2.jpg), (**b**) probe bus schematic. Black arrows indicate locations of the 4.4 N hydrazine thrusters. Blue arrow indicates spin axis

The five THEMIS probes were identical at launch with 134 kg mass (including 49 kg of hydrazine monopropellant). Each measures approximately $0.8 \times 0.8 \times 1.0$ meters (Harvey et al. 2008). On orbit, each has deployed a number of instrument booms and is spin-stabilized at $\sim$20 RPM. Figure 3(a) shows a THEMIS probe with booms deployed. Figure 3(b) shows a schematic of the bus design. The blue arrow, which indicates the spin vector, shall be referred to as the probe $+Z$ direction.

Each probe has four thrusters, nominally 4.4 N each, with locations indicated by the black arrows in Fig. 3(b). Two provide axial thrust (acceleration in $+Z$ direction) for large ΔV maneuvers and attitude control. The other two provide tangential thrust in the spin plane for small ΔV maneuvers and spin rate control. Note that the probes cannot apply acceleration in the $-Z$ direction. During the nominal THEMIS mission, P1 and P2 were flown with the $-Z$ axis close to the ecliptic north pole, i.e., in an "upside-down" configuration relative to ecliptic north and opposite the inner three probes. This was done to aid the main orbit correction maneuvers in the second year of THEMIS, which were designed to counteract lunar perturbations on the orbit plane (Frey et al. 2008). ARTEMIS would maintain the same orientation, as it is quite fuel-intensive to impart spin-axis changes to the probes. Thus, maneuvers towards ecliptic north could not be included in the ARTEMIS trajectory design. At launch, each probe had 960 m/s total ΔV capability (Harvey et al. 2008). At the start of ARTEMIS maneuvers the remaining ΔV (approximately 320 m/s for P1 and 467 m/s for P2) were available for the ARTEMIS trajectory design. Due to fuel tank depressurization (Sholl et al. 2007; Frey et al. 2008), each thruster is expected to produce between 2.4 N and 1.6 N force during the ARTEMIS mission.

Because the spacecraft is spinning the effective thrust of a sideways burn is further reduced, so a maneuver in a particular direction in the spin plane is performed by pulsing the thrusters on and off during each revolution. With a 60 deg pulse duration , the thrusters are on only one-sixth of the time (16.7% duty cycle). Because thrusters are swinging through an arc, the thrust in the desired direction is further reduced to 95.5% effective thrust; with a 40 deg duty cycle the thrusters average only one-ninth thrust, but lose only 2% in efficiency averaged through the arc of each pulse. Only the second reduction in each case influences the effective I_{sp}, so a 40 deg duty cycle would be preferred to a 60 deg one except that lower thrust means longer burns during periapse passages, which would increase gravity losses.

The thermal and power systems have been designed to withstand shadowing from the Sun for up to three hours (Harvey et al. 2008). It was demonstrated in March of 2009, however, that a 4-hour shadow is survivable with appropriate precautions. This limit is therefore being used as the maximum allowable shadow duration for the ARTEMIS mission design, where "shadow" is defined to be less than 50% sunlight.

3 ARTEMIS Concept Development

The baseline THEMIS mission design included the expectation that P1 would experience inordinately long (>8 hr) shadows by March 2010. Although the apoapse altitude of the P1 orbit could have been reduced to prevent this, THEMIS scientists and JPL mission designers came up with the idea of sending P1 "up" instead of "down" in 2005. With THEMIS instrumentation, compelling science could be conducted near or at the Moon with a single probe. According to initial trajectory studies, a direct transfer from P1 Earth orbit to a 1500 km altitude by 18,000 km radius polar orbit at the Moon would require ~500 m/s of ΔV (not including margin or losses associated with long thrust arcs). This was well beyond P1's expected ΔV capability at the end of the baseline mission.However, the remaining fuel appeared sufficient to transfer P1 from its Earth orbit to the desired eccentric lunar orbit by way of a lunar swing-by and low-energy transfer (Chung et al. 2005). When initiated by a lunar swing-by, this type of transfer does not require any less ΔV to leave Earth, but saves essentially all the ΔV cost of getting into a Lissajous orbit around one of the Earth-Moon Lagrange points. It does this by using solar gravity tidal perturbations to make the three-body energy change on the trajectory that would otherwise have to be done propulsively at arrival near the Moon. The fuel reserves on P2 offered similar capability, suggesting the possibility of sending two THEMIS probes to the Moon.

With the encouraging initial trajectory design results in hand, proposals for funding to support a detailed design study of low-energy trans-lunar trajectories, feasibility studies related to the THEMIS hardware, and optimization of the remaining THEMIS mission for P1 and P2 were made in 2006 and 2007. Although these proposals were not selected for funding, the science team continued concept development as time permitted.

In the summer of 2007, internal JPL funding became available to support an Explorer program Mission of Opportunity proposal for a THEMIS mission extension that would become ARTEMIS. A team from the JPL Inner Planets Mission Analysis group was convened to design trajectories to the Moon for P1 and P2. Building on the work done in 2005, the JPL team (working closely with the THEMIS science and mission operations teams) developed a workable trajectory within THEMIS probe constraints that provided the opportunity for a highly rewarding scientific mission. This formed the baseline trajectory of the current ARTEMIS mission. Midway through this preliminary design effort, NASA headquarters advised the ARTEMIS team that the new mission would be more appropriately proposed as an extended mission for THEMIS, rather than as a mission of opportunity. At around the same time, the mission operations team at UCB-SSL was augmented by navigators and maneuver designers at GSFC who contributed operations experience with Lissajous and translunar orbits to the design effort.

The complete preliminary design for the extended mission was presented to the Senior Review Board for the Heliophysics Division in February 2008 (Angelopoulos and Sibeck 2008); approval to proceed with detailed design was given in May of that year. The preliminary trajectory design that was presented to the Senior Review Board is described in

Broschart et al. (2009). This paper is an update of that earlier design paper; the design described there has changed significantly since the approval to proceed. As was understood at the time, the series of Earth orbits leading up to the initial lunar flybys needed to be significantly redesigned. More recently, a number of changes have been made in the science operations phase of ARTEMIS.

In 2009 it was recognized that significant additional scientific benefits from ARTEMIS could be obtained for the Planetary Division of NASA's Science Mission Directorate. The team was invited to propose an amendment to its Heliophysics plan that addressed Planetary objectives. The proposal was returned by NASA/HQ, and the invitation was re-extended for submission in the 2010 Senior Review cycle, so both Heliophysics and Planetary aspects of the ARTEMIS proposal could be evaluated by a joint panel. ARTEMIS/Heliophysics was given the go-ahead to continue operations in June 2010. The ARTEMIS/Planetary decision, though delayed until December 2010, was also positive. The 2008 preliminary design of ARTEMIS's lunar orbits needed to be modified to accommodate planetary objectives by lowering periapse altitudes, raising inclinations, and adjusting the lines of apsides for better overlap of measurements with those of NASA's Lunar Atmosphere and Dust Environment Explorer (LADEE) mission.

The Planetary Division's decision to execute the planetary objectives of the mission came only 3 months prior to the baseline ARTEMIS lunar orbit insertion (originally slated for April 2011). This did not leave sufficient time for performing the necessary lunar orbit optimization to meet the expanded science objectives. Therefore, the team decided to postpone insertion to June-July 2011 to enable further study of the planetary aspects of the investigation. This postponement in turn entailed modifications to both the Lissajous phase and the transition to lunar orbits.

The ARTEMIS science objectives and the characteristics of orbits that would satisfy them (for both Heliophysics and Planetary Divisions of the Science Mission Directorate) as proposed and accepted by the 2010 Senior Review were described in Angelopoulos (2010). The revised mission design described in this paper represents the ARTEMIS orbit execution plan as actually implemented.

4 ARTEMIS Science Goals

Angelopoulos (2010) gives a comprehensive overview of ARTEMIS mission science objectives and describes how the mission design and operations are structured to meet them. Here we describe aspects of the mission that drive mission design.

Each probe is equipped with a suite of five particle and field instruments used to study geomagnetic substorm activity during the nominal THEMIS mission. These instruments include a Fluxgate Magnetometer, a Search Coil Magnetometer, an Electric Field Instrument, an Electrostatic Analyzer, and a Solid State Telescope (Angelopoulos 2008). This instrumentation suite allows the probe to measure the 3D distribution of thermal and superthermal ions and electrons and the AC and DC magnetic and electric fields to study the interaction between the Earth's magnetic field and the Sun's magnetic field and solar wind. By expanding the spatial extent of THEMIS's multiple, identically-instrumented spacecraft, ARTEMIS allows us to study plasmoids in the magnetotail, particle acceleration and turbulence in the magnetotail and the solar wind. Furthermore, ARTEMIS will study lunar wake formation and evolution for the first time with two identical, nearby probes, thereby resolving spatio-temporal ambiguities. The aforementioned heliophysics objectives of the mission can be addressed by inter-spacecraft separations and wake downstream crossings that are initially as large as 20 Earth radii and are progressively reduced to 1000 km or less. This goal is achieved initially by having the ARTEMIS probes at large separations in Lissajous orbits around two (and later one) of the Earth-Moon Lagrange points, and subsequently by insertion of probes into lunar orbits with $\sim$ 18,000 km apoapse radius and highly variable angular separation between their lines of apsides.

ARTEMIS also offers a unique opportunity to contribute to planetary science. From its unique orbits ARTEMIS will study the "sources and transport of exospheric and sputtered species; charging and circulation of dust by electric fields; structure and composition of the lunar interior by electromagnetic (EM) sounding; and surface properties and planetary history, as evidenced in crustal magnetism. Additionally, ARTEMIS's goals and instrumentation complement LRO's *[Lunar Reconnaissance Orbiter's]* extended phase measurements of the lunar exosphere and of the lunar radiation environment by providing high fidelity local solar wind data. ARTEMIS's electric field and plasma data also support LADEE's prime goal of understanding exospheric neutral particle and dust particle generation and transport" (Angelopoulos 2010).

To achieve these objectives, ARTEMIS requires both high- and low-altitude measurements by one spacecraft, while the other measures the pristine solar wind nearby. Low periapses are very important in increasing the ability of ARTEMIS to measure sputtered ions and crustal magnetism in situ. For this reason periapse altitudes less than 50 km are highly desired. Additionally, the latitude of periapsis is an important consideration for lunar crustal magnetism—increased periapsis latitude provides opportunities for covering a larger portion of the lunar surface. A latitude greater than 10 deg (goal 20 deg) is highly desirable. Finally, conjunctions with LADEE at the dawn terminator necessitate that one of the ARTEMIS probes have its periapsis positioned near the dawn terminator and pass through periapse close to the time of LADEE passage through that region. These design considerations have been incorporated into the current planning for the upcoming lunar orbit insertions (**LOI**s).

5 ARTEMIS Trajectory Design

Figure 1 shows the ARTEMIS trajectory design that sent P1 and P2 from their respective orbits at the end of the THEMIS primary mission to insertion into lunar Lissajous orbit. The P1 trajectory is shown in red, and the P2 trajectory is shown in blue. The design succeeded in meeting both the trajectory constraints imposed by the probe capabilities and the requirements derived from the science objectives.

In the following subsections, the trajectory is broken up into phases for detailed discussion. These include the Earth orbit phase, the trans-lunar phase, the Lissajous orbit phase, and the lunar orbit phase. An integrated timeline of the events for P1 and P2 in these four mission phases can be found in Table 1; an integrated ΔV budget is given in Table 2.

5.1 Earth Orbit Phase Trajectories

When the preliminary design was being developed to show the feasibility of ARTEMIS, the orbit raise did not appear to present any particular challenge, so this phase was simplified to a single impulsive velocity increase at perigee, followed by a number of Earth orbits including lunar approaches that modified the orbit and culminated in the lunar flyby that begins the low-energy transfer to the Moon. This simplification allowed one track of the design effort to focus most strongly on the lunar flyby and transfer; the series of finite orbit raise maneuvers (ORMs) to raise the Earth orbit could be developed later in parallel on a separate design track.

Table 1 Integrated trajectory design timeline

Earth-Orbit Phase	Jul 20, 2009	ARTEMIS mission begins
Earth-Orbit Phase	Jul 21, 2009	First P2 Orbit-Raise Maneuver
Earth-Orbit Phase	Aug 1, 2009	First P1 Orbit-Raise Maneuver
Earth-Orbit Phase	Oct 12, 2009	P1 Fly-by Targeting Maneuver 1
Earth-Orbit Phase	Dec 2, 2009	P1 Fly-by Targeting Maneuver 2
Earth-Orbit Phase/ Trans-Lunar Phase	Jan 31, 2010	P1 Lunar Fly-by #1 (min Range = 12600 km)
Trans-Lunar Phase	Feb 13, 2010	P1 Lunar Fly-by #2 (min Range = 3290 km)
Trans-Lunar Phase	Mar 10, 2010	P1 Deep-space Maneuver (+ Local Maximum Range = 1200000 km to Earth)
Earth-Orbit Phase	Mar 24, 2010	P2 Fly-by Targeting Maneuver
Earth-Orbit Phase/ Trans-Lunar Phase	Mar 28, 2010	P2 Lunar Fly-by (min Range = 8070 km)
Trans-Lunar Phase	Apr 13, 2010	P1 Earth Fly-by (min Range = 17000 km)
Trans-Lunar Phase	May 11, 2010	P2 Earth Fly-by #1 (min Range = 86000 km)
Trans-Lunar Phase	Jun 1, 2010	P2 Deep-space Maneuver 2
Trans-Lunar Phase	Jun 06, 2010	P1 Maximum Range (1,500,000 km to Earth)
Trans-Lunar Phase	Jun 18, 2010	P2 Maximum Range (1,200,000 km to Earth)
Trans-Lunar Phase	Jul 27, 2010	P2 Earth Fly-by #2 (min Range = 170000 km)
Trans-Lunar Phase/ Lissajous Orbit Phase	Aug 25, 2010	P1 LL2 Insertion
Trans-Lunar Phase/ Lissajous Orbit Phase	Oct 20, 2010	P2 LL1 Insertion
Lissajous Orbit Phase	Jan 01, 2011	P1 Departs LL2
Lissajous Orbit Phase	Jan 11, 2011	P1 LL1 Insertion
Lissajous Orbit Phase	Jun 18, 2011	P1 Lunar Transfer Initiation
Lissajous Orbit Phase	Jun 21, 2011	P2 Lunar Transfer Initiation 1
Lunar Orbit Phase	Jun 27, 2011	P1 LOI (1850 km alt)
Lissajous Orbit Phase	Jun 28, 2011	P2 Lunar Transfer Initiation 2
Lunar Orbit Phase	Jul 17, 2011	P2 LOI (3800 km alt)
Lunar Orbit Phase	Dec 28, 2012	P1 End of 1.5 year Lunar Orbit Phase
Lunar Orbit Phase	Jan 17, 2013	P2 End of 1.5 year Lunar Orbit Phase
LADEE Science Phase	Jul 7, 2013	Beginning, for earliest LADEE launch
LADEE Science Phase	Oct 15, 2013	End, for earliest LADEE launch
LADEE Science Phase	Dec 16, 2013	Beginning, for latest LADEE launch
LADEE Science Phase	Mar 26, 2014	End, for latest LADEE launch

Figure 4 shows the ARTEMIS P1 trajectory from the end of the nominal THEMIS mission through the first close lunar flyby. In the figure, the red line represents the ARTEMIS P1 trajectory starting with its orbit at the end of the THEMIS primary mission, and the gray circle indicates the Moon's orbit. The plot is centered on the Earth and shown in the Sun-Earth synodic coordinate frame, which rotates such that the Sun is fixed along the negative X axis (to the left) and the Z axis is aligned with the angular momentum of the Earth's heliocentric orbit. As time passes, the line of apsides of P1's geocentric orbit rotates clockwise in the main figure. The insert in the bottom left shows P1's motion out of the ecliptic plane, where

Table 2 ARTEMIS ΔV budget as proposed and actual (with italic values showing current estimates of lunar deorbit ΔV)

	P1 cost est. (m/s)	P1 cost act. (m/s)	P2 cost est. (m/s)	P2 cost act. (m/s)
ORMs	96.7	95.8	204.0	231.4
SDMs				11.0
FTMs	7.0	6.9	5.7	12.4
DSMs	4.8	7.3	15.1	30.2
LTI	1.5	3.2	0.8	1.0
LOIs	89.9	50.3	117.1	73.0
Lunar orbit periapse lowering		40.7		45.3
Deterministic DV total	200	204	343	404
Sources of additional DV cost:				
TLI declination penalty	(Included)	(Included)	(Included)	(Included)
TLI grav and steering loss (w/shadow)	(Included)	(Included)	36	(Included)
LOI declination penalty	2	(Included)	2	(Included)
LOI grav and steering loss	(Included)	(Included)	(Included)	(Included)
Lissajous maintenance	15	8.7	12	4.5
TCMs (3% + 1 m/s per ORM $\times \sqrt(n)$)	15	7.5	14	4.2
Total	232	221	407	413
Available DV	324	320	475	467
Margin	92	99	68	54
Liens against margin:				
Matching ORM phase to transfer phase	None		5	
Precession correction in ORM phase	1		2	
Lissajous maintenance increase	20		13	
End-of-mission deorbit	10	*2*	64	*2*

the largest plane change was caused by a lunar approach in December 2009. The labels on the plot provide information about key events during this phase of the mission.

The design of the P2 Earth orbits phase was similar, as shown in Fig. 5, but lasted two months longer because it started from a smaller Earth orbit and a longer series of finite maneuvers needed to be included to raise the orbit.

As we gradually came to realize, the reference trajectory design for the Earth orbit phase of both P1 and P2 would turn out to be significantly more complex than a simple series of maneuvers to replace the preliminary design's impulsive orbit raise maneuver. This complexity stemmed from: (1) probe operational constraints, (2) the tight ΔV budget, (3) the precision phasing required to reach the designed low-energy transfers to the Moon, and (4) the actual initial states for ARTEMIS P1, P2 in the summer of 2009. These actual states ended up significantly different from the initial states that were predicted in 2005–2007; this change was due to deterministic orbit-change maneuvers that occurred in 2008, mid-way through the THEMIS mission, to improve science yield for the second THEMIS tail season (Fig. 6 shows this difference for the P1 orbit). As expected, the actual orbit raise required perigee burns on multiple orbits due to the small thrust capability. The design of these burns was challenging because generally an optimal design of highly elliptical transfers is numerically difficult, and because lunar approaches created a complex three-body design space.

Rotating Sun-Earth Coordinates, In-Plane View
Lunar Approach:
8-Dec-2009 @ 18800 km
Lunar Approaches:
17-Sep-2009 @ 45700 km
14-Oct-2009 @ 131000 km
Lunar Fly-by #1:
31-Jan-2010 @
12600 km altitude
To Sun
Orbit Raise Maneuvers (ORMs):
1-Aug-2009 – 13-Sep-2009
5 (10) maneuvers, 95.8 m/s total
Fly-by Targeting Maneuvers
(FTMs):
0.8 m/s on 12-Oct-2009
6.1 m/s on 2-Dec-2009
Moon's Orbit
Trajectory Correction
Maneuvers (TCMs):
1.9 m/s on 14-Dec-2009
1.5 m/s on 15-Jan-2010
0.3 m/s on 24-Jan-2010
Out-of-Plane
End of Nominal Mission Orbit:
8240 x 202000 km (1.29 x 31.6 RE),
3.93 day period
Total Time from ORM-1 to 1st Fly-by: 183 days
Orbit periods vary from 3.93 to 11.3 days
Moon's semi-major axis: 384000 km (60RE)

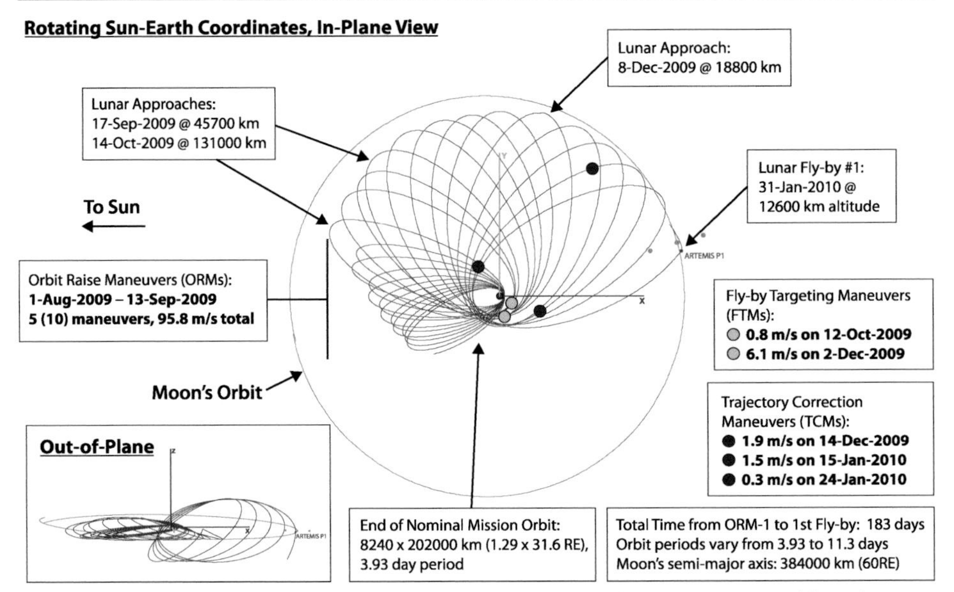

Fig. 4 Earth orbit portion of the P1 trajectory design. Distances quoted are ranges measured from the center of mass of the Earth or Moon

Rotating Sun-Earth Coordinates, In-Plane View
Fly-by Targeting Maneuver (FTM):
12.4 m/s on 24-Mar-2010
Trajectory Correction Maneuver (TCM):
0.6 m/s on 26-Mar-2010
Shadow Deflection
Maneuvers (SDMs):
3.6 m/s on 17-Nov-2009
7.4 m/s on 2-Dec-2009
Lunar Approach:
1-Feb-2010 @ 70600 km
To Sun
Lunar Approach:
1-Mar-2010 @ 68700 km
Orbit Raise Maneuvers (ORMs):
21-Jul-2009 – 26-Feb-2010
27 (39) maneuvers, 231.4 m/s
total
Lunar Fly-by:
28-Mar-2010 @
8070 km altitude
Out-of-Plane
End of Nominal Orbit:
9710 x 124000 km (1.52 x 19.4 RE),
1.98 day period
Total Time from ORM-1 to 1st Fly-by: 250 days
Orbit periods vary from 1.98 to 10.0 days
Moon's semi-major axis: 384000 km (60RE)

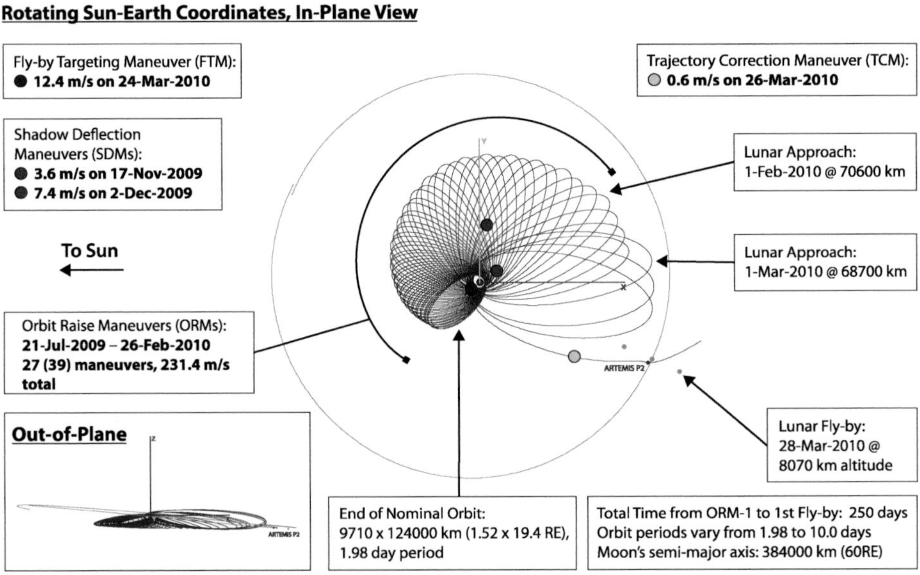

Fig. 5 Earth orbit portion of the P2 trajectory design. Distances quoted are ranges measured from the center of mass of the Earth or Moon

During the refinement of the orbit design, it was recognized that several factors conspired to further complicate the development of the reference trajectory:

1. Earth's shadow covers perigee for much of the orbit raise season, prohibiting thrusting at/near perigee. The design necessitated splitting most perigee burns into two (A and B) burn arcs bracketing the shadow, further increasing burn arc length and gravity losses.

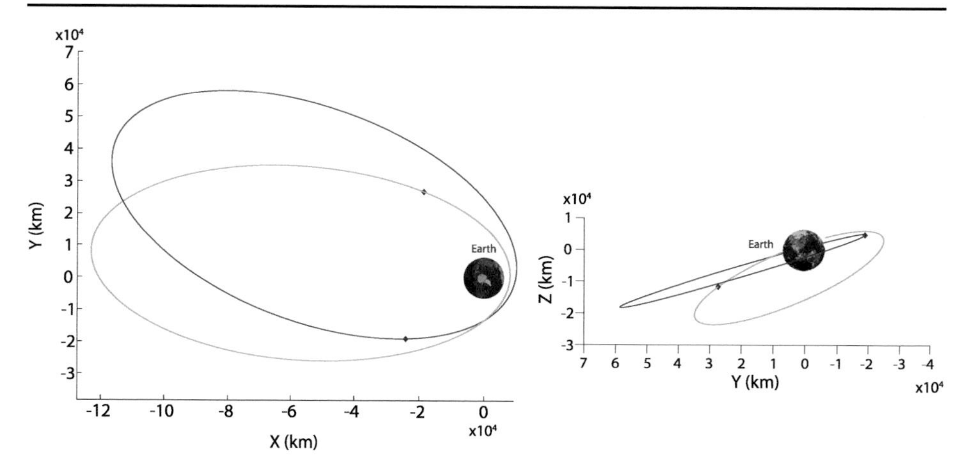

Fig. 6 (**a**) Initial orbit of the Earth orbit portion of the P1 preliminary trajectory design. The initial condition for ARTEMIS P1 predicted when ARTEMIS was proposed is in *green*; the actual starting orbit is in *red*. (**b**) End-on view of (**a**)

2. The initial propellant load of $\sim 50\%$ for P2 forced a large fraction of the maneuvers to be performed at a lower duty cycle (shorter pulse) due to the propellant load being near a "slosh resonance" (Sholl et al. 2007; Auslander et al. 2008; Frey et al. 2008). This further exacerbated gravity losses, necessitating more maneuvers to obtain the same total orbit-raise ΔV. This was addressed by starting the ORM sequence for P2 as early as July 20, 2009.
3. Side thrusting for orbit-raise maneuvers also results in a small reorientation (precession) of the spin axis due to a small offset of the thrust direction relative to the probe center of mass. The cumulative effect of side thrusting has been significant spin-plane precession of the probes in directions that either violated operational constraints or increased losses from vector-thrusting. Spin axis reorientation maneuvers were included in the mission design to account for that effect.
4. Thrust restrictions due to the absence of "up" thrusting capability posed a non-traditional restriction to the mission design. The usual intuition that 1 burn allows targeting of 3 elements and 2 burns separated in time allows for the targeting of 6 elements is not correct for ARTEMIS. In fact, even 3 separated burns can fail to provide 6-element targeting when all maneuvers are confined to a single plane.

5.1.1 Orbit-Raise Design Process

The P1 and P2 orbit-raise designs were constructed using Mystic software (Whiffen 1999; Whiffen 2006). Mystic was able to accommodate all mission constraints outlined above. However, the complex (and often treacherous) design space resulting from numerous lunar approaches during the orbit-raise phase made simple design strategies impossible. To provide some robustness against missed burns, and sufficient tracking data for orbit/maneuver reconstruction, perigee maneuvers were double-spaced, i.e., two orbits apart. On occasion it proved advantageous to separate burns even farther to take advantage of or avoid strong lunar interactions. Most perigee burns were divided into and modeled as two separate burn arcs, one on either side of the Earth's shadow. The duration and pointing of each burn was fully optimized using Mystic, with the constraint that the end states of this phase would be on the translunar trajectories already designed.

Several different end-to-end orbit-raise strategies were thus attempted for both P1 and P2, with the desired translunar injection as a goal and the initial ARTEMIS state as a starting point as early as needed, i.e., with an ascend start date unrestricted by THEMIS science considerations. The strategy that proved most successful for the P1 trajectory was to first optimize sets of burns on three double-spaced perigees to reach an orbital period of 131 hours. From states near this point forward, there existed a tremendous number of possible paths involving differing lunar interactions, numbers of Earth revolutions, plane changes, and node changes over the next 140 days of ballistic propagation. It was not at all obvious which of these many paths might be feasible, and then which feasible path would be best to rejoin the low-energy transfer. To address this problem, a large number of ballistic trajectories were used as initial guesses for targeting and optimization. Different families were organized based on the number of Earth revolutions. A computer cluster was used for this compute-intensive process. Trajectories that were found to be feasible or nearly feasible were then further refined by moving the time of rejoining the low-energy transfer to successively later dates.

5.1.2 P1 and P2 Orbit-Raise Designs

The P1 low-energy transfer began with a pair of lunar flybys separated by only 14 days—see Figs. 7 and 8. To minimize the ΔV cost of getting onto the designed translunar trajectory, it seems desirable to match these flybys as closely as possible, though exact matching does not seem to be necessary. Intuitively, re-joining the low-energy transfer at later times would provide increasing efficiency, since a longer time would allow a lower rendezvous velocity. It was expected (and found) that re-joining much beyond the second lunar flyby provided diminishing returns. The final total effective ΔV for P1's Earth orbit phase as actually flown was 102.7 m/s (compare this to the 103.7 m/s allocation (see Table 2) and the 125 m/s conservative estimate in the ARTEMIS proposal (Angelopoulos and Sibeck 2008) from a single-impulse Earth departure, which included 24 m/s for gravity and steering losses and trajectory correction maneuvers (**TCM**s)). The final design maneuvers are given in Table 3,

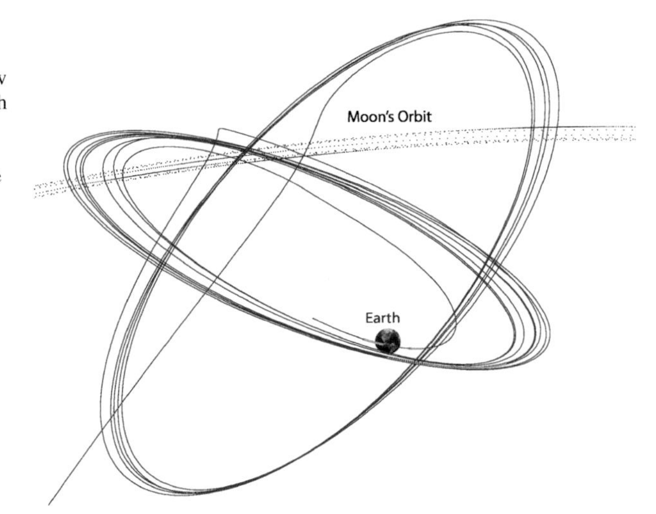

Fig. 7 P1's final Earth orbits leading into the low-energy transfer, showing an oblique view of the final lunar approach (which changed the orbit inclination significantly) and the first of the two lunar flybys that initiated the transfer. The *green arcs* along two of the orbit periapses show where the flyby targeting maneuvers (FTM1A, FTM1B, and FTM2) were performed

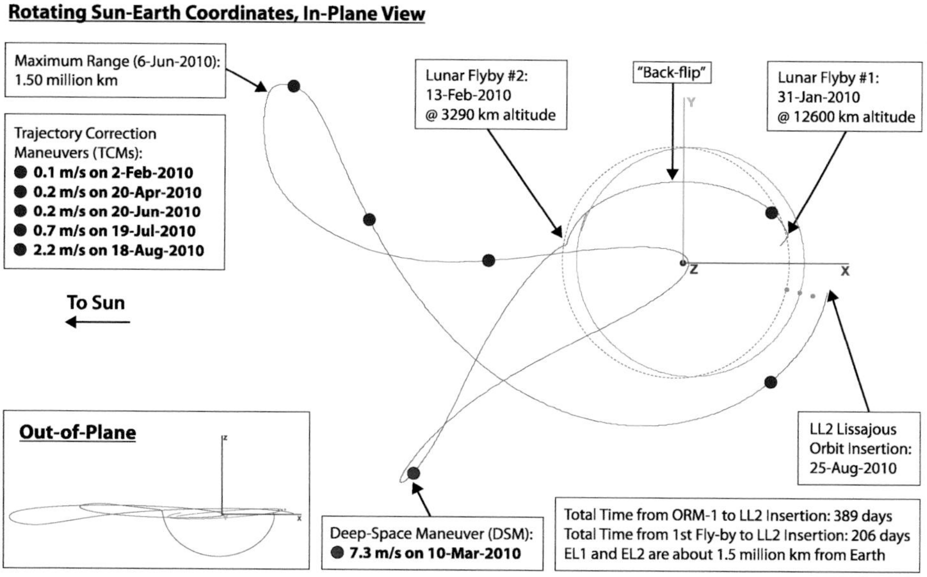

Fig. 8 Trans-lunar portion of the P1 trajectory design. Distances quoted are ranges measured from the center of mass of the Earth or Moon

along with trajectory correction maneuvers designed by the mission operations team during the execution of this phase.

The P2 orbit design was more complex than the P1 design because P2 begins in a much smaller orbit. A process similar to the P1 design process was used to develop the P2 orbit-raise design. Very careful planning of distant lunar approaches was necessary to stay within the allocated ΔV budget, which was more constraining for P2 than for P1. The P2 orbit raise required 42 burns, counting each split maneuver as two burns (see Table 4). The method used was a branching process. Each orbit-raise maneuver was designed several times to reach different orbital periods (different period = different "branch"). Subsequent maneuvers reaching longer periods were designed for each branch. The most promising branches were continued; poorly performing branches were abandoned. Poorly performing branches often led to situations in which lunar interactions reduced the orbit period or required long periods without maneuvers to avoid disadvantageous lunar interactions. Highly performing branches ended up with advantageous distant lunar interactions early on. Distant lunar interactions that provided maneuver savings as little as 1 meter per second early in the orbit raise were sought. The final few orbit-raise maneuvers required very careful planning to maximize the positive influence of the Moon.

A major additional complication of the P2 trajectory design occurred shortly before the first ORM, when a check for eclipses found an unacceptably long passage through Earth's shadow just after the ORMs and before the first lunar flyby. Additional shadow-deflection maneuvers (SDMs) were added to change the orbit plane to reduce the time in shadow and then change the orbit plane back to return to the planned flyby conditions. These SDMs solved the problem without requiring a complete redesign of the series of ORMs, though at a cost of 11 m/s in additional ΔV. Even with these maneuvers added, the final total effective ΔV for P2's Earth orbit phase as actually flown was 254.8 m/s (compare to the 245.7 m/s allocated and the 219 m/s originally estimated (Angelopoulos and Sibeck 2008) from the

Table 3 P1 Earth orbit and transfer phase maneuvers

ORM1A	2009/213	20:01:04.853	8.243
ORM1B	2009/213	20:50:51.171	8.441
ORM2A	2009/222	13:01:15.050	8.389
ORM2B	2009/222	13:51:33.867	8.468
ORM3A	2009/232	07:06:30.159	8.477
ORM3B	2009/232	07:56:36.098	8.517
ORM4A	2009/243	08:12:29.465	13.901
ORM4B	2009/243	09:14:21.187	11.855
ORM5A	2009/256	18:57:06.519	13.984
ORM5B	2009/256	19:48:19.402	5.505
FTM1A	2009/285	08:38:00.635	0.203
FTM1B	2009/285	08:41:51.037	0.602
FTM2	2009/336	08:02:21.215	6.084
TCM1	2009/348	04:51:56.825	1.886
TCM2	2010/015	12:27:38.304	1.455
TCM3	2010/024	07:00:59.591	0.311
TCM4	2010/033	07:10:53.521	0.116
DSM1	2010/069	19:00:00.000	7.312
TCM5	2010/110	09:00:00.000	0.179
TCM6	2010/171	21:45:00.000	0.180
TCM7	2010/200	23:00:00.000	0.651
TCM8	2010/230	06:00:00.000	2.244

single-impulse Earth departure, which included 33 m/s for gravity and steering losses and TCMs). The final design and trajectory correction maneuvers are given in Table 4.

5.2 Trans-lunar Phase Trajectories

The trans-lunar phase of the ARTEMIS trajectory for each probe extended from the first close lunar flyby to insertion into the target Lissajous orbit.

Figure 8 shows the trans-lunar phase of the ARTEMIS trajectory for P1. The trajectory is shown in the same Sun-Earth synodic coordinate frame used in Figs. 4 and 5. In the figure the trajectory begins on the right side of the plot with "Lunar Fly-by #1". The P1 trajectory made use of a "back-flip", wherein the first lunar fly-by set up a second lunar fly-by on the opposite side of the Moon's orbit ~14 days later. The back-flip can be seen clearly in the out-of-plane view insert in the bottom left of Fig. 8 and the beginning of it is shown in Fig. 7. This second flyby raised the apogee significantly, throwing the probe out beyond the Moon's orbit towards the Sun. This began the low-energy trajectory leg for P1, which is characterized by significant gravitational perturbation imparted on the probe by the Sun. This low-energy trajectory had two deep-space legs that included one relatively small deep-space maneuver (DSM). After the second leg, the orbit perigee had been raised to lunar distance, and the phasing with the Moon's orbit was such that the probe moved into a lunar Lissajous orbit around lunar Lagrange point #2 (LL2) without requiring any deterministic insertion maneuver. By the time P1 reached the Lissajous orbit in August of 2010, 389 days had elapsed since the start of ARTEMIS maneuver operations.

Figure 9 shows the trans-lunar trajectory for P2. The P2 trajectory only included one lunar fly-by, which sent the probe away from the Sun and beyond the Moon's orbit into a

Table 4 P2 Earth orbit and transfer phase maneuvers

ORM1	2009/202	07:33:03.552	10.686
ORM2	2009/206	10:41:40.407	5.292
ORM3A	2009/210	15:10:44.501	2.399
ORM3B	2009/210	16:10:50.116	8.348
ORM4A	2009/215	00:46:49.019	3.324
ORM4B	2009/215	01:47:36.747	8.695
ORM5A	2009/219	15:24:58.115	3.870
ORM5B	2009/219	16:22:13.277	7.915
ORM6A	2009/224	11:22:18.944	3.903
ORM6B	2009/224	12:16:13.013	6.952
ORM7A	2009/229	12:35:08.258	3.867
ORM7B	2009/229	13:26:36.035	6.257
ORM8A	2009/234	19:27:11.635	4.267
ORM8B	2009/234	20:18:08.362	6.058
ORM9A	2009/240	08:07:08.656	3.871
ORM9B	2009/240	08:53:50.173	4.753
ORM10A	2009/246	02:16:11.659	4.729
ORM10B	2009/246	03:01:38.535	4.387
ORM11A	2009/252	02:39:19.201	5.000
ORM11B	2009/252	03:21:47.856	3.419
ORM12A	2009/258	09:02:17.385	5.586
ORM12B	2009/258	09:43:04.698	2.831
ORM13A	2009/264	22:31:20.114	5.882
ORM13B	2009/264	23:11:39.983	2.507
ORM14A	2009/271	18:34:32.874	7.092
ORM14B	2009/271	19:14:07.763	2.240
ORM15	2009/278	23:21:22.389	5.881
ORM16	2009/286	09:49:58.506	8.599
ORM17	2009/294	06:15:08.345	9.851
ORM18	2009/302	13:41:36.289	10.269
ORM19	2009/311	10:47:56.571	10.113
ORM20	2009/320	22:41:23.999	10.039
ORM21	2009/331	01:50:58.997	4.148
ORM22	2009/341	12:28:02.607	2.083
ORM23	2009/352	07:22:49.060	4.782
ORM24	2009/363	11:22:37.941	6.233
ORM25	2010/010	05:05:57.689	7.580
ORM26	2010/022	19:08:17.058	5.846
ORM27	2010/057	08:52:20.815	11.875
SDM1	2010/059	08:17:18.815	3.636
SDM2	2010/074	09:55:42.965	7.360
FTM1	2010/083	16:07:17.000	12.406
TCM1	2010/085	02:05:41.282	0.648
DSM1	2010/133	02:21:16.534	3.685
DSM2	2010/152	14:50:00.000	23.280
TCM2	2010/201	12:00:00.000	2.152
TCM3A	2010/214	11:58:28.326	0.634
TCM3B	2010/214	12:01:37.625	0.090
DSM3A	2010/252	13:58:24.982	2.478
DSM3B	2010/252	14:01:59.393	0.797
TCM4	2010/274	11:00:00.000	0.306
TCM5	2010/285	13:40:00.000	0.250

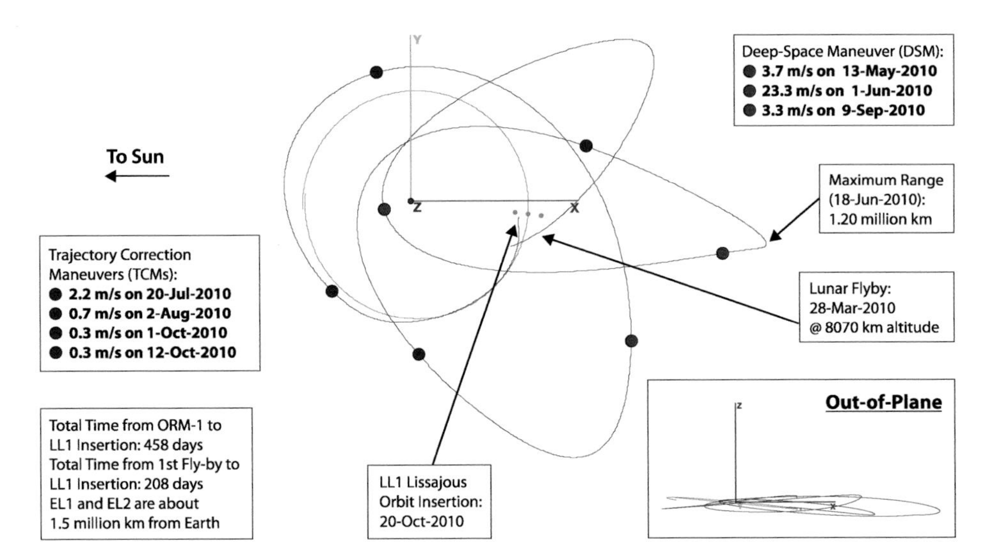

Fig. 9 Trans-lunar portion of the P2 trajectory design. Distances quoted are ranges measured from the center of mass of the Earth or Moon

region where the perturbative influence of solar gravity is significant. P2 followed a low-energy trajectory that included three deep-space legs before entering a lunar Lissajous orbit around lunar Lagrange Point #1 (LL1) without any deterministic thrusting. The P2 trajectory included three deep-space maneuvers (DSM), one of which was relatively large; these maneuvers totaled 30.4 m/s. P2 arrived in Lissajous orbit about 2 months after P1, requiring a total of 458 days since the start of ARTEMIS maneuver operations to reach this stage.

5.2.1 Transfer Trajectory Implementation

As the transfer trajectory was flown, correction maneuvers were required to adjust for earlier maneuver execution and probe pointing and implementation errors, as well as navigation errors. These maneuvers, called trajectory correction maneuvers (TCMs), encompassed the statistical maneuvers along the transfer. TCMs in addition to DSMs were inserted in each of the P1 and P2 designs.

We allocated 4% of the total propellant budget of each probe to perform any required TCMs along the way to control the energy to keep P1 and P2 near their appropriate outgoing trajectories. Since the two probes had already completed their primary mission in a highly elliptical Earth orbit, propellant was extremely limited. Thus, with the unique operational constraints, accomplishment of the transfer goals with the minimum propellant cost was the highest priority. To implement the mission design, our trajectory simulations use a full ephemeris model with point-mass gravity representing Earth, Moon, Sun, Jupiter, Saturn, Venus, and Mars. Also included is an eighth degree and order Earth potential model. The solar radiation pressure force is based on: (1) the measured probe area, (2) the probe estimated mass (from bookkeeping), and (3) the coefficient of reflectivity determined by navigation estimation. The same models with estimates for the mass usage and coefficient of reflectivity were used in the mission design process that determined the reference trajectory. The trajectory propagations in operations were based on a variable step Runge-Kutta

8/9 and Prince-Dormand 8/9 integrator. Initial conditions used throughout the planning process correspond to the UCB-delivered navigation solutions using the DSN and the UCB tracking system. Software tools used in this process include the General Mission Analysis Tool (GMAT) developed at GSFC as an open source, high-fidelity tool with optimization and MATLAB connectivity and AGI's STK/Astrogator suite.

To compute actual commanded maneuver ΔV requirements, we used two numerical methods: differential corrections (DC) targeting using central or forward differencing and an optimization method using the VF13AD algorithm from the Harwell library. A DC process provided *a priori* conditions. Equality constraints were incorporated for DC application; nonlinear equality and inequality constraints were employed for optimization. These constraints incorporated both the desired target conditions in the Earth-Moon system and probe constraints on the ΔV direction and relationship between the spin axis and the ΔV vector.

The end goal of the transfer phase was to achieve the Earth-Moon Lissajous insertion conditions necessary for a minimal energy insertion into the Earth-Moon L2 or L1 Lissajous orbits. The goals were defined in terms of states expressed in Earth J2000 coordinates. These targets were held constant over the entire mission design and implementation process once the reference translunar transfer had been designed. Although a baseline trajectory was defined to design the mission, the adaptive strategy used in operations required exactly matching this baseline only at the end of the transfer.

5.2.2 Navigation Uncertainties

Throughout the transfer trajectory implementation process, navigation solutions were generated at a regular frequency of once every three days with the exception of post-maneuver navigation solutions, which were made available as soon as a converged solution was determined. The rapid response was to ensure that the maneuver had performed as predicted and that no unanticipated major changes to the design were necessary. The RSS of the uncertainties were on the order of tens of meters in position and below 1 cm/s in velocity. As a conservative estimate for maneuver planning and error analysis, 1σ uncertainties of 1 km in position and 1 cm/s in velocity were used. These accuracies were obtained using nominal tracking arcs of one three-hour contact every other day. The Goddard Trajectory Determination System (GTDS) was used for all navigation estimations.

5.2.3 Trajectory Design During Operations

The transfer trajectory implementation approach used the numerical methods discussed above augmented by dynamical systems theory for verification and to gain knowledge of the transfer dynamics. The probes were targeted to the libration point orbit insertion locations knowing full well that maneuver execution and navigation errors would push the path off the "baseline" design. A correction maneuver scenario was planned that would essentially shift the trajectory, such that the new path would be consistent with a nearby manifold. It was decided to use a forward-integrating numerical optimization process that included probe constraints to calculate optimized ΔVs. This procedure permitted minimization of the ΔV magnitude, variation of the ΔV components in direction, as well as variation of the maneuver epoch, while incorporating the nonlinear constraint on the probe ΔV direction relative to the spin axis.

Originally, it was envisioned that errors in navigation and maneuvers could lead to the need for an unobtainable correction in an "up" direction with respect to the ecliptic plane. Fortunately, experience with trajectory design on other missions that incorporate weak stability regions near Sun-Earth libration orbits and near the ecliptic plane showed us that we could allow upward ΔV corrections to be delayed until an equivalent magnitude but opposite direction (downward) ΔV location could be found in the long-duration transfer. These locations were then used to correct the trajectories without any upward maneuver component to achieve the final Earth-moon insertion targets.

As the TCMs were performed, the path essentially jumped from the vicinity of one local transfer manifold to another at a slightly different energy level. The number of optimized TCMs was very low and their magnitudes quite small, considering the sensitivity of the dynamics and uncertainties of the OD solutions.

5.2.4 *Maneuver Design*

To target to the desired Earth-Moon Lissajous conditions, a VF13AD optimizer was used. We optimized each maneuver to determine the minimal ΔV location. To determine an *a priori* maneuver location and to achieve an intuitive feel for the maneuver results, a DC process was first performed using planned DSN coverage. For P1, the first four TCMs were completed in Earth-centered elliptical orbit or during lunar gravity-assist targeting. Maneuver execution errors are small, only a few percent. These errors are a function of actual start time with respect to a sun pulse of a spinning spacecraft, tank temperatures, attitude knowledge, and general propulsion system performance.

It should be noted that maneuver execution errors, current navigation errors, and subsequent maneuvers to correct for these errors along with small mis-modeled perturbations can lead not only to late or early arrival times at the prescribed Lissajous insertion location, but also may contribute to out-of-plane effects and may result in trajectories that intersect with the Moon. Clearly, the trajectory is very sensitive to such small variations. But that sensitivity also implies that small corrections can alter the trajectory design significantly and allow low ΔV cost orbit control, assuming sufficiently frequent tracking for orbit reconstruction.

5.3 Lissajous Orbit Phase Trajectories

The Lissajous orbit phase of ARTEMIS has permitted repeated observations of the distant lunar wake. For the first ~1.5 months of this phase (from August 22 to October 2, 2010), P1 was alone at the Moon in orbit around the LL2 point while P2 was still en route. P2 then arrived, making a partial orbit around LL2 on its way to Lissajous orbit at LL1. For about the next 2.3 months, P1 orbited LL2 while P2 orbited LL1, and then P1 also crossed over to orbit LL1. During this phase, the trajectories permit 16 independent observations of the lunar wake when crossing behind the Moon on the anti-Sun side, observations of the distant Earth magnetotail once per month when the Moon's orbit passes through it, and observations of the pristine solar wind when out of the influence of both. These two-point measurements were made at separation scales up to ~100,000 km when the probes were in orbit around different Lagrange points and up to ~50,000 km when both orbit LL1. Distant magnetotail measurements can also be correlated with concurrent measurements from THEMIS-A, THEMIS-D, and THEMIS-E in low-Earth orbit.

Figure 10 shows the P1 trajectory during the Lissajous orbit phase. In this figure, the Moon is at the origin and the trajectory is drawn in the Earth-Moon synodic coordinate frame, which rotates such that the Earth is always to the left along the negative X axis. The Z axis is aligned with the angular momentum vector of the Moon's geocentric orbit. The main figure on the left side shows the view looking down on the geocentric orbital plane of the Moon, and the two insets show perspectives from within the Moon's orbital plane. The LL1 and LL2 points are marked in the figure.

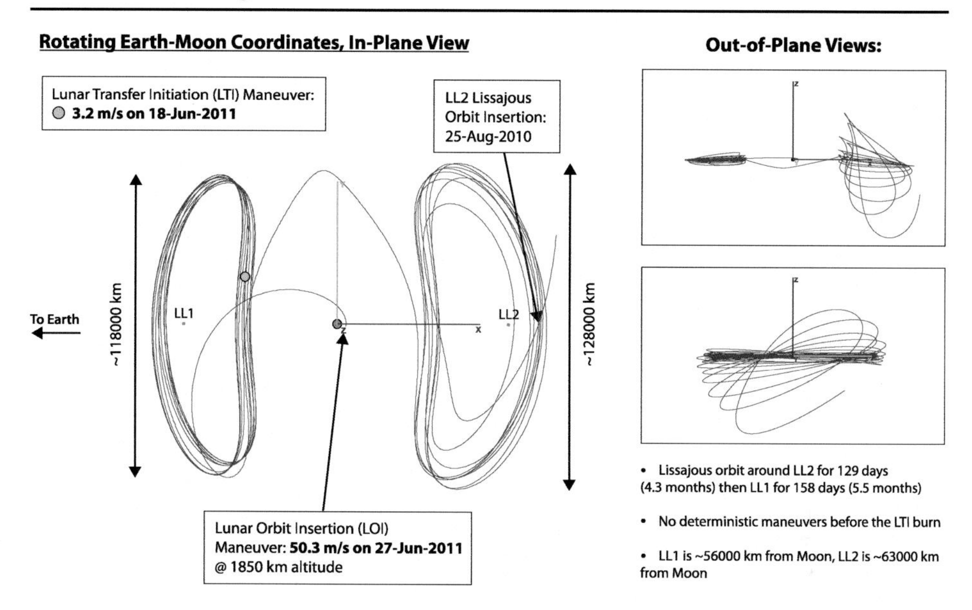

Fig. 10 Lissajous orbit phase of the P1 trajectory. Distances given are ranges measured from the lunar center of mass unless otherwise specified

P1 entered Lissajous orbit around LL2 on August 25, 2010 with only a stochastic maneuver to begin station-keeping. Although the initial Lissajous orbit was somewhat inclined with respect to the Moon's geocentric orbit plane, the orbit flattened after a few orbits (see Fig. 10 inserts). After $\sim$129 days in orbit around LL2, P1's trajectory followed an unstable orbit manifold along a 10-day heteroclinic connection to a Lissajous orbit around LL1 (Howell et al. 1997; Koon et al. 2000). Although this transfer required no deterministic ΔV for initiation or insertion, in practice weekly station-keeping maneuvers (SKMs) were required to maintain the Lissajous orbit. P1 spent 158 days orbiting LL1 before executing a small maneuver to depart from Lissajous orbit on June 18, 2011. The probe descended to an 1850 km periselene altitude, where the lunar-orbit insertion (LOI) maneuver was executed, beginning the lunar orbit phase on June 27, 2011. At the time of LOI, P1 had operated for 707 days since the beginning of the ARTEMIS mission.

Figure 11 shows the P2 trajectory during the Lissajous orbit phase. P2 entered Lissajous orbit around LL1 on October 20, 2010. As with P1, this insertion was achieved without any deterministic ΔV because the incoming trans-lunar trajectory approached on the stable manifold of this particular Lissajous orbit. P2 stayed in this nearly planar Lissajous orbit for about 8.5 months before initiating descent to a $\sim$3800 km altitude periselene. The LOI maneuver for P2 occurred on July 17, 2011, at which time P2 had been operating for 727 days since the end of the nominal THEMIS mission.

After P1 and P2 entered their Lissajous orbits, a project decision was made to extend the Lissajous phase from April to July. This required adding axial components to SKM18 on February 1, 2011, for P1 and to SKM11 (January 4), SKM13 (January 18), and SKM15 (February 1) for P2. These axial burns, which directly affected the Z velocity of the probes in the Earth-Moon frame, were needed to prevent the Z axis components of the Lissajous orbit states from oscillating too much. These oscillations otherwise would have grown to uncontrollable levels before the transition to lunar orbits despite ARTEMIS's stationkeeping process.

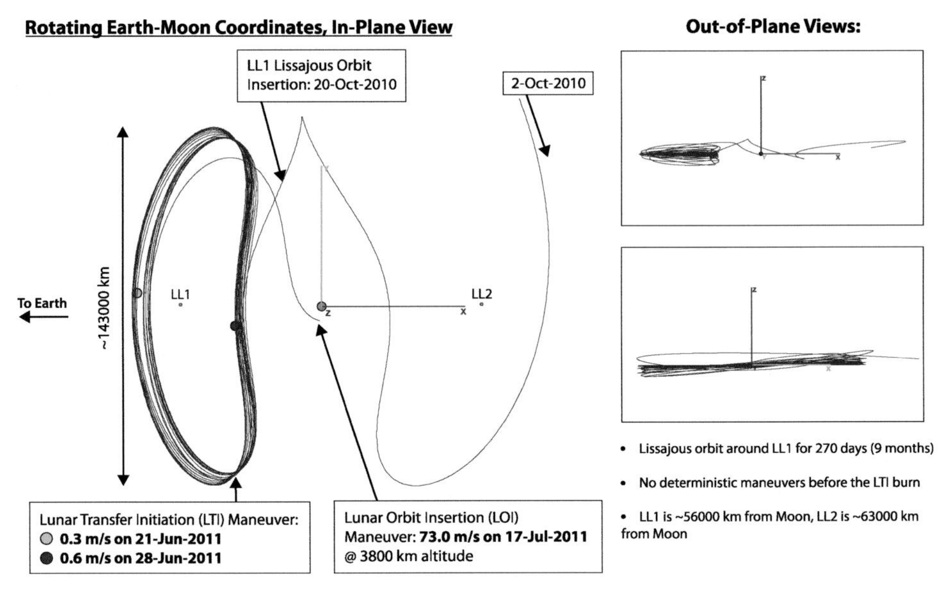

Fig. 11 Lissajous orbit phase of the P2 trajectory design. Distances given are ranges measured from the lunar center of mass unless otherwise specified

The Lissajous orbit phase of the ARTEMIS mission were particularly exciting because the ARTEMIS probes are the first to fly in a lunar Lissajous orbit. Flying these orbits was a challenge for operations and maneuver design teams because Lissajous orbits are inherently very unstable; small, unavoidable deviations from the Lissajous orbit are amplified to problematic proportions (Howell and Keeter 1995) after approximately one revolution ($\sim$14 days). This leaves little room for error in the operations. Because of this instability, correction maneuvers needed to be executed about weekly to keep the probes in orbit. So even though these orbits required no deterministic ΔV, orbit maintenance ΔV was required.

5.3.1 Stationkeeping

There are many stationkeeping methods to chose from: classical control theory or linear approximations of Farquhar (1971), Farquhar and Kamel (1973), Breakwell and Brown (1979), and Hoffman (1993), provided analysis and discussion of stability and control in the Earth-Moon collinear L1 and L2 regions; Renault and Scheeres (2003) offered a statistical analysis approach; Howell and Keeter (1995) addressed the use of selected maneuvers to eliminate the unstable modes associated with a reference orbit; and Gómez et al. (1998) developed and applied the approach specifically to translunar libration point orbits. Folta et al. (2010) presented an analysis of stationkeeping options and transfers between the Earth-Moon locations and the use of numerical models that include discrete linear quadratic regulators and differential correctors.

The ARTEMIS stationkeeping method used maneuvers performed at optimal locations to minimize the ΔV requirements while ensuring continuation of the orbit over several revolutions downstream. There are no reference trajectories to plan against, so other methods such as linear (continuous) controllers are impractical. Likewise, other targeting along the X axis or Y axis is more costly or cannot be attained without violating probe constraints. Goals in the form of energy achieved, velocities, or time at any location along the orbit can be used, but our goal was defined in terms of the X velocity component at the X axis crossings. This assumes selection of a velocity that can be related to the orbit energy at any particular time. To initialize the analysis, a DC scheme is used, based on the construction of an invertible sensitivity matrix by numerical sampling of orbital parameters downstream as a consequence of specific initial velocity perturbations (Folta et al. 2010). The orbit is continued over several revolutions by checking the conditions at each successive goal then continued to the next goal. This allows perturbations to be modeled over multiple revolutions.

The targeting algorithm uses an impulsive maneuver with variables of either Cartesian ΔV components or ΔV magnitude and azimuth angle within the ARTEMIS spin plane. Target goals are specified uniquely for each controlled orbit class because LL1 and LL2 dynamics differ slightly. The velocity target chosen is specifically set to continue the orbit in the proper direction. Targeting is then implemented with parameters assigned at the X–Z plane crossing such that the orbit is balanced and another revolution is achieved. Each impulsive maneuver is targeted to the X component of the velocity at the third X axis crossing after the maneuver; the maneuver supplies velocity (energy) in a direction that subsequently continues the libration point orbit. Additionally, the VF13AD1 optimizer is used to minimize the stationkeeping ΔV by optimizing the direction of the ΔV and the location (or time) of the maneuver. Included in the DC and optimization process are constraints required to keep the ARTEMIS maneuvers in the spin plane.

Given the constraints of the ARTEMIS mission orbit, probe maneuvers were planned at a seven-day frequency to ensure a stable navigation solution while minimizing the ΔVs and staying within the ARTEMIS ΔV budget. The maneuvers were originally planned to occur at or near the X axis crossings and to use a continuation method to maintain the orbit. As operational experience was gained, however, it was found possible to relax the location of each maneuver in order to permit a more user-friendly operational schedule. Orbital conditions were set to permit the energy or velocity at the crossings to continue the orbit for at least 2 revolutions.

Since oscillations in the Z component of the state are largely decoupled from motion in the X–Y plane and are not as unstable as that motion, our expectation was that these oscillations could be controlled using maneuvers only in the probes' $+Z$ directions, which were close to ecliptic south for each probe; this proved to be the case for this oscillation control (described earlier in this section) as well as for statistical Z control throughout the Lissajous phase. The complete set of stationkeeping maneuvers for both probes is detailed in Tables 5 and 6.

5.4 Lunar Orbit Phase Trajectories

Most scientific observations for ARTEMIS occur during the lunar orbit phase, which nominally lasts 2 years. It was desirable for the lunar orbit to have apoapsis as high as possible in order to enable a large range of downstream lunar wake measurements. Another goal for the mission was to maximize the number of periapses less than 50 km altitude in order to best measure crustal magnetism and sputtered ions. Long orbit lifetimes and at least some inclination relative to the lunar equator would help to maximize the variety of measurement opportunities. Considering also the spacecraft capabilities, the ARTEMIS lunar orbits were chosen to have apoapsis radius around 19,000 km (driven by the maximum acceptable eclipse duration) with periapsis altitudes varying between roughly 20 and 1200 km altitude. The P1 orbit is retrograde with a lunar periapsis latitude range of ± 12 deg and the P2 orbit is prograde with a periapsis latitude range of ± 17 deg.

Table 5 P1 Lissajous maneuvers

SKM1	2010/237	04:30:00.000	2.562
SKM2	2010/251	11:00:00.000	0.584
SKM3	2010/265	09:00:00.000	0.223
SKM4	2010/273	16:25:00.000	0.341
SKM5	2010/282	16:30:00.000	0.078
SKM6	2010/291	14:00:00.000	0.158
SKM7	2010/298	07:00:00.000	0.113
SKM8	2010/306	05:00:00.000	0.116
SKM9	2010/313	01:45:00.000	0.066
SKM10	2010/321	08:45:00.000	0.072
SKM11	2010/334	05:55:00.000	0.210
SKM12	2010/344	06:30:00.000	0.227
SKM13	2010/352	14:30:00.000	0.139
SKM14	2010/361	17:15:00.000	0.117
SKM15	2011/006	18:40:00.000	0.033
SKM16	2011/017	06:55:00.000	0.120
SKM17	2011/024	08:00:00.000	0.062
SKM18A	2011/032	18:35:00.000	2.100
SKM18B	2011/032	18:45:00.000	0.192
SKM19	2011/038	19:05:00.000	0.224
SKM20	2011/045	06:10:00.000	0.104
SKM21	2011/049	20:45:00.000	0.014
SKM22	2011/056	04:20:00.000	0.060
SKM23	2011/063	00:00:00.000	0.030
SKM24	2011/070	03:45:00.000	0.029
SKM25	2011/076	10:25:00.000	0.017
SKM26	2011/083	02:35:00.000	0.023
SKM27	2011/089	19:10:00.000	0.020
SKM28	2011/096	18:55:00.000	0.019
SKM29	2011/103	10:30:00.000	0.022
SKM30	2011/110	02:00:00.000	0.281
SKM31	2011/116	16:45:00.000	0.029
SKM32	2011/124	14:15:00.000	0.130
SKM33	2011/131	07:20:00.000	0.061
SKM34	2011/144	15:55:00.000	0.056
SKM35	2011/150	20:40:00.000	0.012
SKM36	2011/157	19:40:00.000	0.043
SKM37/LTI	2011/169	00:31:00.000	3.229
SKM38/LTI-TCM1	2011/173	01:00:00.000	0.509

The conic orbit elements of these lunar orbits are subject to constant change primarily induced by Earth's perturbing gravitational influence during the high apoapses. In the lunar orbit phase, periapsis altitudes vary by 800–1000 km twice per lunar orbit around the Earth (i.e., roughly every two weeks) due to tidal forces. Because the Earth's location relative to the Moon's surface is nearly fixed, this two week oscillation in periapsis altitude always has a minimum near the same lunar longitudes: 90 deg and 270 deg for a retrograde orbit (P1)

Table 6 P2 Lissajous phase maneuvers

SKM1	2010/293	12:50:00.000	0.117
SKM2	2010/300	06:15:00.000	0.184
SKM3	2010/307	14:05:00.000	0.379
SKM4	2010/315	10:50:00.000	0.247
SKM5	2010/322	05:25:00.000	0.063
SKM6	2010/333	04:45:00.000	0.350
SKM7	2010/340	22:55:00.000	0.104
SKM8	2010/348	03:40:00.000	0.066
SKM9	2010/355	12:40:00.000	0.036
SKM10	2010/362	16:25:00.000	0.122
SKM11	2011/004	16:42:00.000	0.960
SKM12	2011/011	20:40:00.000	0.120
SKM13	2011/018	13:45:00.000	0.170
SKM14	2011/025	10:05:00.000	0.177
SKM15A	2011/032	02:10:00.000	0.030
SKM15B	2011/032	02:15:00.000	0.038
SKM16	2011/039	08:35:00.000	0.294
SKM17	2011/050	02:20:00.000	0.173
SKM18	2011/057	17:00:00.000	0.036
SKM19	2011/065	05:50:00.000	0.216
SKM20	2011/072	22:30:00.000	0.211
SKM21	2011/079	10:45:00.000	0.044
SKM22	2011/086	20:30:00.000	0.020
SKM23	2011/100	20:10:00.000	0.050
SKM24	2011/107	10:55:00.000	0.045
SKM25	2011/116	10:25:00.000	0.017
SKM26	2011/123	17:30:00.000	0.068
SKM27	2011/130	18:35:00.000	0.023
SKM28	2011/138	02:05:00.000	0.019
SKM29	2011/143	17:05:00.000	0.014
SKM30	2011/151	00:05:00.000	0.024
SKM31	2011/161	04:00:00.000	0.068
LTI1	2011/172	01:00:00.000	0.347
LTI2	2011/179	16:15:00.000	0.622
LTI-TCM	2011/186	06:20:00.000	0.086

and 0 deg and 180 deg for a prograde orbit (P2). Another longer term eccentricity oscillation is induced by the secular precession of the argument of periapsis caused by the Earth for the ARTEMIS orbits (Scheeres et al. 2001). This oscillation causes periapsis altitudes to vary by a few hundred km every few months. The P1 and P2 periapsis altitudes through 2013 clearly show the influence of these two oscillations (Fig. 12).

Another effect of Earth's perturbation on the orbits is to cause the ecliptic longitude of the periapse for each orbit to change in the same direction as the orbital motion by about 100 deg per year. By putting the probes into opposing orbits, e.g., P2 prograde and P1 retrograde, the relative motion of their lines of apsides is maximized. The combination of this apsidal motion with the significant eccentricity of the orbits enables observations to be achieved at

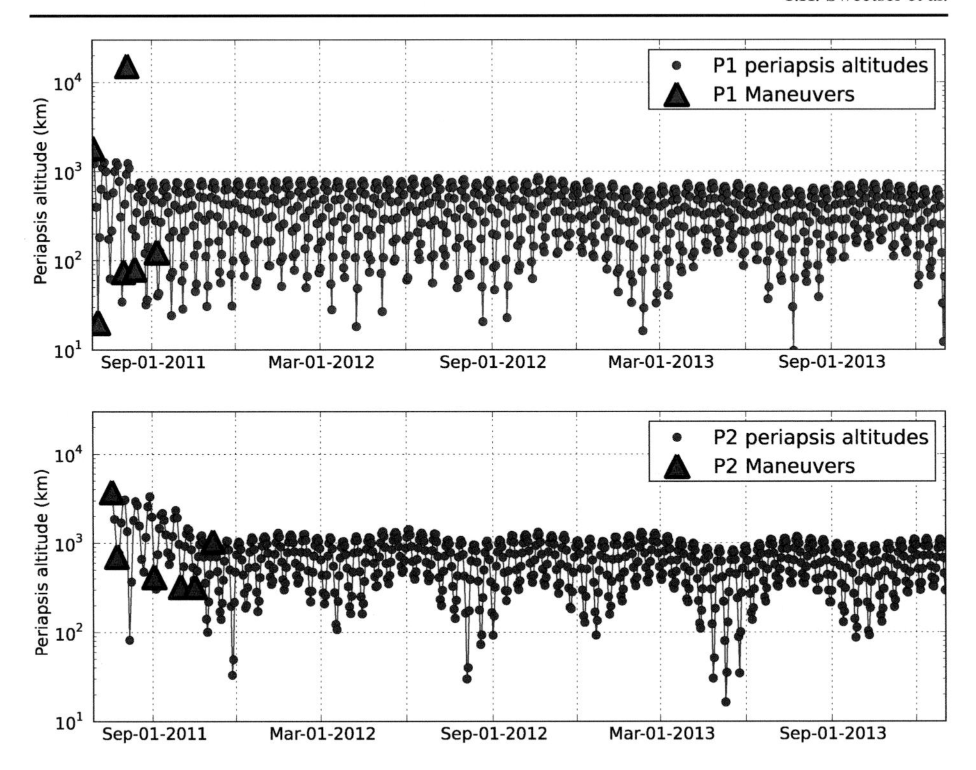

Fig. 12 Altitude at periapse for P1 and P2 in lunar orbit. The bi-weekly and every-few-monthly oscillations in periapsis altitude are clearly shown. Note that small PEB maneuvers are planned for 2012 and 2013 but are not explicitly labeled here

a wide range of probe separations (from ~150 to ~30,000 km) and geometries in the lunar wake.

5.4.1 Lunar Orbit Insertion and Achieving the Science Orbit

Achieving the ARTEMIS science orbits was a challenging problem due to the small size of the tangential thrusters, the limited ΔV available, and the aforementioned strong Earth perturbations. The approach trajectories from Lissajous orbit could not enter the science orbit with a single lunar orbit insertion (LOI) maneuver because not enough ΔV was available. It was necessary to divide the insertion into many maneuvers to reduce the gravity and steering losses to an acceptable level. The implemented transfer design consists of an LOI maneuver plus five period reduction maneuvers (PRMs) for each spacecraft. The geometries of the P1 and P2 approaches from Lissajous and subsequent transfers to the science orbit are shown in Fig. 13. The timing and ΔV of each maneuver, as executed, is given in Table 7.

The first steps toward achieving the science orbit were performed during the Lissajous orbit phase. An out-of-plane component was included in SKMs in January and February of 2011 that set the inclination of approach to LOI such that an acceptable science orbit inclination could be achieved. This method of inclination modification is very fuel efficient relative to a plane-change maneuver in lunar orbit.

Relatively large LOI maneuvers with significant gravity losses were required to capture P1 and P2 into a low enough orbit so that Earth-gravity perturbations did not result in impact

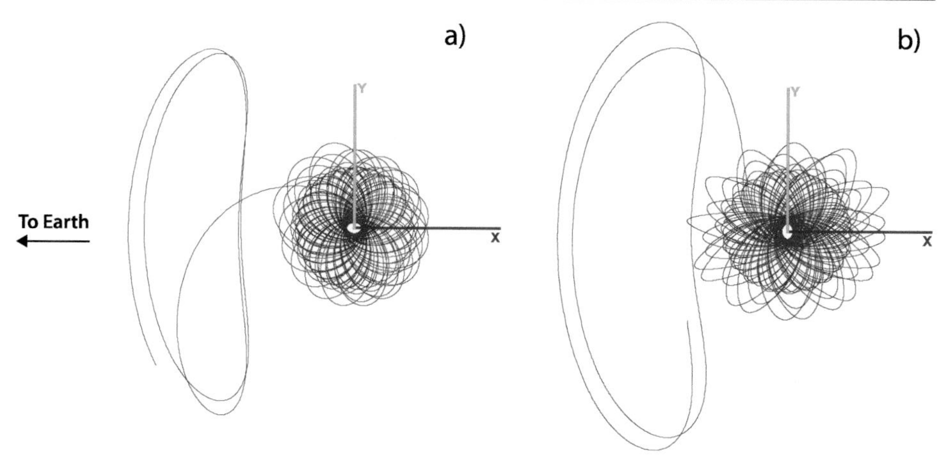

Fig. 13 LOI and low-lunar orbit trajectories for (**a**) P1 and (**b**) P2 in the rotating Moon-centered frame. LL1 Lissajous orbits shown for scale

Table 7 LOI and PRM maneuver details for both spacecraft in chronological order. It was an important design criterion that the activities on one spacecraft did not interfere with planning or execution of activities on the other. The total ΔV expended on these maneuvers was 91.1 m/s for P1 and 118.4 m/s for P2

Burn name	Date	Total ΔV (m/s)	# of segments
P1 LOI	June 27, 2011	50.3	3
P1 PRM-1	July 3, 2011	12.1	2
P2 LOI	July 17–18, 2011	73.0	3
P2 PRM-1	July 23, 2011	8.5	2
P1 PRM-2	July 31, 2011	4.6	1
P1 PRM-3	August 3, 2011	2.5	1
P1 PRM-4	August 12, 2011	13.0	2
P2 PRM-2	September 2, 2011	12.4	2
P1 PRM-5	September 7, 2011	8.6	1
P2 PRM-3	October 1, 2011	13.3	2
P2 PRM-4	October 17, 2011	7.7	2
P2 PRM-5	November 7, 2011	3.5	1

on subsequent periapses. The geometry to the approach from Lissajous orbit determines the initial phasing on the bi-weekly periapsis altitude oscillation and the magnitude of the oscillation grows with the apoapsis altitude. The approach phase was effectively fixed due to dynamics and the future objective to coordinating measurements with LADEE. Thus, to avoid impact at the minimum of the oscillation cycle, the LOI altitude had to be high enough and LOI duration long enough to avoid a surface-impacting post-LOI orbit. The P1 LOI altitude was selected to be 1850 km and the P2 LOI altitude was at 3800 km. At these altitudes, the minimum safe burn duration for LOI was 135 minutes for P1 and 173 minutes for P2 with the roughly 0.5 N thrust available at the time of LOI (assuming use of a ± 30 deg pulse width). The P2 LOI duration was further increased to 205 minutes to improve the eclipse phasing for the subsequent PRMs. Since the ARTEMIS thrust cannot be dynamically steered along an anti-velocity direction without significant redesign of its thruster operations software, the LOIs were divided into three constant-direction thrust segments in order to reduce steering losses; these segments were separated by a minimum of three minutes for operational reasons.

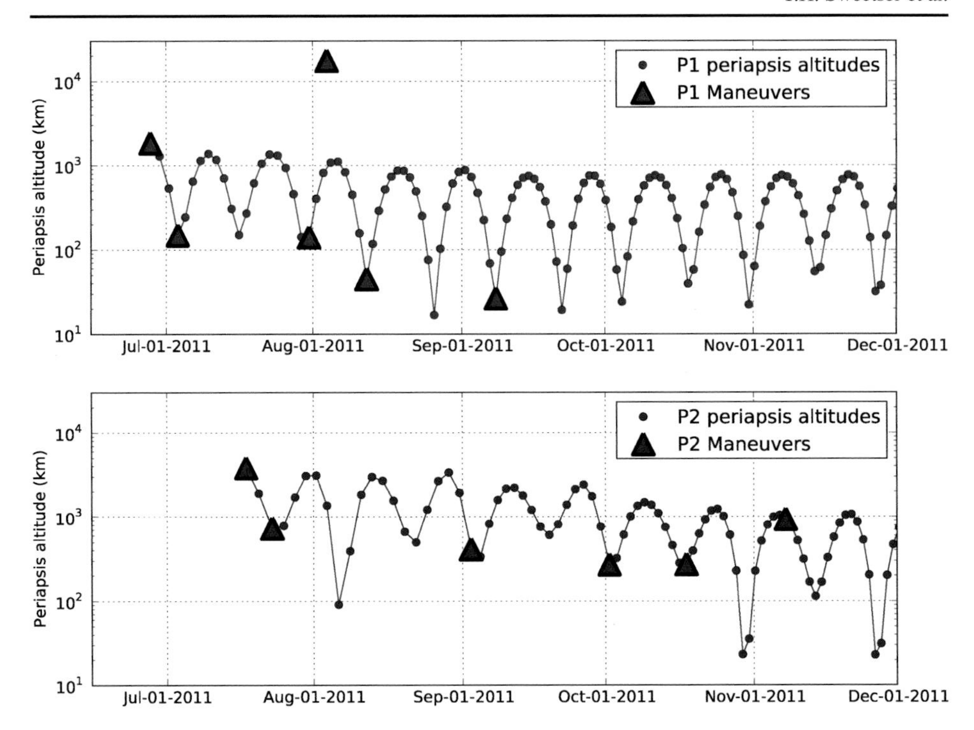

Fig. 14 Time history of periapsis altitudes for P1 and P2 during the first 6 months in orbit. LOI and PRM maneuver dates are marked as *triangles*. All maneuvers were roughly centered around periapsis with the exception of P1's PRM-3, which was at apoapsis

After a safe capture orbit was achieved with the LOI maneuvers, the rest of the transition to the science orbits was achieved with a series of smaller and more efficient PRM burns. For maximum fuel-efficiency, an infinite number of very small PRMs at periapses would be ideal. For ARTEMIS, a total of 10 PRMs (5 for each spacecraft) were performed over a period of 5 months. PRMs were included on periapse numbers 3, 20, 28, and 49 and apoapsis number 22 for P1. For P2, PRMs were placed at periapse numbers 2, 20, 36, 47, and 63. The placement of these PRM maneuvers relative to the bi-weekly periapsis altitude oscillation is shown in Fig. 14. A number of factors were considered in determining the placement and sizing of the PRM maneuvers including: ΔV efficiency (using the lowest periapsis altitudes is most efficient), operations schedule (sufficient time for orbit determination, maneuver design, and testing must exist between maneuvers), operations staff availability (P1 and P2 activities should not be simultaneous since a single team operates both spacecraft), orbit lifetime (should not be less than 14 days if a maneuver is missed), final stable orbit altitude (PRMs needed to manage periapsis altitudes to ensure a low altitude final orbit), eclipses (spacecraft cannot maneuver in eclipse and shadow times must be less that 4 hrs), occultations (flight rules prohibit maneuver initialization when out of radio contact), and the performance of preceding PRMs (the PRM sequence design was modified as needed after each maneuver execution). The total characteristic ΔV expended for execution of the combined LOI and PRMs was 91.1 m/s for P1 and 118.4 m/s for P2, where *characteristic* ΔV is the ΔV the thrusters would have provided if the spacecraft weren't spinning. Additional detail on the design of the transfer from Lissajous to the science orbit can be found in Broschart et al. (2011).

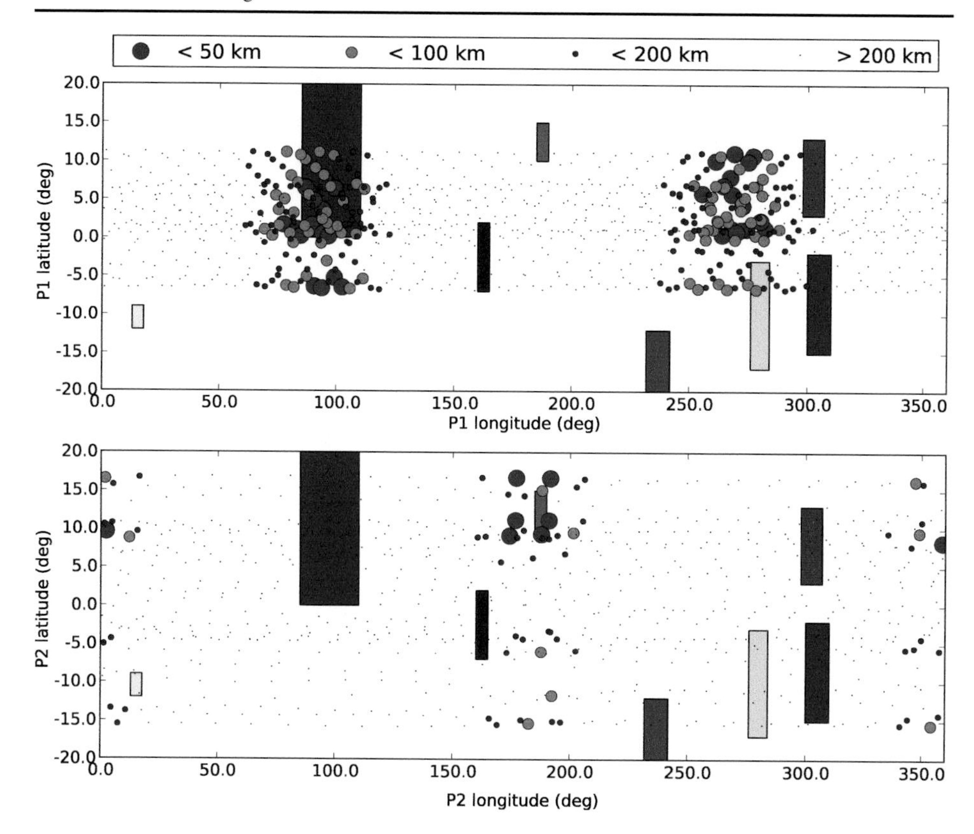

Fig. 15 Periapse locations at the Moon during four years of lunar orbits. Periapse altitudes are affected by perturbations due to Earth's gravity so that they are lowest at constrained longitudes for P1 and P2. The lunar orbits have been tuned to give good coverage of certain magnetic anomalies in the lunar crust (shown as shaded rectangles)

5.4.2 Science measurement opportunities

Figure 12 gives current predictions for the periapsis altitudes that are expected through the end of 2013. One of the most challenging aspects of the transfer to science orbit was maximizing the number of sub-50 km periapses subject to mission and spacecraft constraints. The science orbits that have been successfully achieved are such that the natural dynamics allow for a number of low periapsis opportunities. The absence of significant secular drift in the periapsis altitude helps to minimize the maintenance ΔV needed in the coming years to avoid impact. Planetary-science enhancement burns (PEBs) are planned (but not finalized) to maintain the periapsis altitude as needed and, ultimately, to lower the apoapsis to increase the number of periapses and reduce the altitude oscillation magnitude.

The science orbits were inclined from the lunar equator, and the periapses were, in part, driven to be so low to allow for better measurements of crustal magnetic anomalies. Figure 15 shows the periapsis altitude and locations for both spacecraft relative to the lunar surface. The colored boxes indicate the location of some known crustal magnetic anomalies. The inclination oscillation and the secular movement of the argument of periapsis (Scheeres et al. 2001) induced by the Earth's gravity allow for a range of periapsis latitudes at all longitudes. Some of these periapses offer the opportunity for optimal crustal anomaly mea-

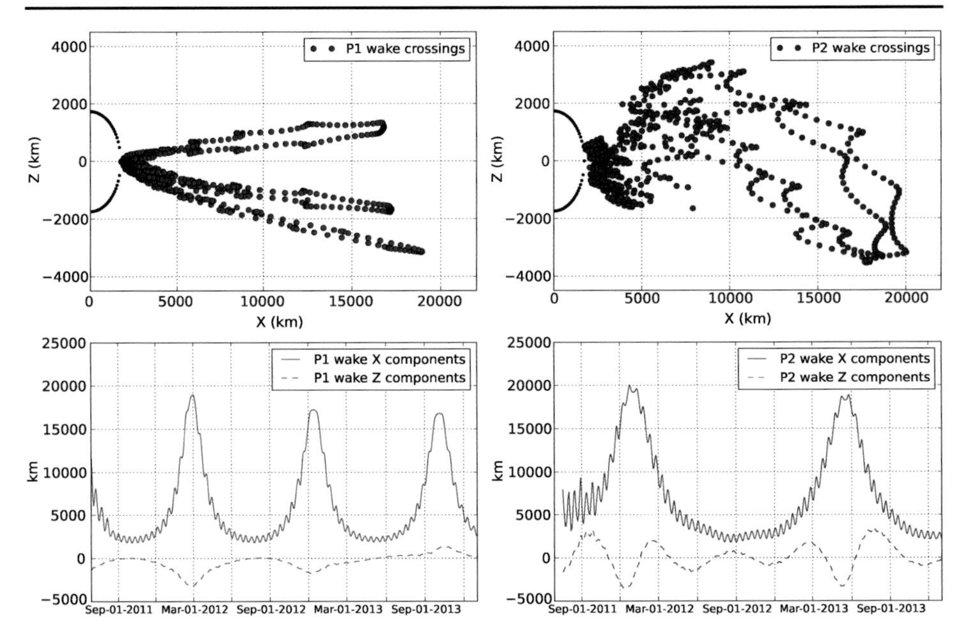

Fig. 16 P1 (*red*) and P2 (*blue*) lunar wake observation opportunities during LOI and the lunar orbit phase, with the Moon's limb indicated by *black dots* and the Sun on the $-X$ axis. Each *red* or *blue point* represents an orbit arc within or near the wake. These arcs vary in length and orientation

surements (i.e., sub-50 km altitude near an anomaly). Note that the dynamics dictate that maximum periapsis latitudes are achieved only when inclination is at the minimum of its oscillation cycle.

A key heliophysics objective of the mission is to measure the lunar wake with the two spacecraft in a number of relative geometries over time. Figure 16 shows the range from the Moon's center in the anti-Sun direction of the lunar wake crossing observation opportunities for the achieved ARTEMIS science orbits as a function of time. A large number of measurements opportunities have been created due to the low orbit inclination (one opportunity per orbit per spacecraft), large variety of down-Sun ranges (due to the orbit eccentricity), and relative geometries (due to the orbit precession induced by the Earth).

6 Mission Status

As of February, 2012, both P1 and P2 have successfully arrived into and maintained Lissajous orbits around the Earth-Moon L1 point, transitioned into lunar orbit insertions, and reduced their orbit periods into the science orbit. Both probes and their instruments are functioning normally. One minor surprise occurred on October 14, 2010, when a small, sudden change was observed in the velocity and spin rate of P1, which was quickly traced to the loss of the EFI sensor ball at the end of one of the four EFI booms deployed from the sides of the probe. This loss was originally attributed to a micrometeorite severing the fine wire that connected the sensor ball to the preamp at the end of the boom (http://www.nasa.gov/mis-sion_pages/themis/news/artemis-struck.html), but then the preamp separated from the same boom on August 27, 2011. This suggests that in the earlier event a micrometeorite impacted and weakened the connection between the boom and the preamp. In this scenario the shock of the impact at the other end of the preamp caused the fine wire to the sensor to break immediately, while thermal cycling and various mechanical strains finally broke the preamp loose later (Owens et al. 2012).

Although reduction in the number of EFI sensors will cause a slight reduction in the quality of the electric field measurements, the instrument still satisfies its science requirements. The loss of the sensor mass also shifted the probe's center of mass, which has and will complicate operations somewhat, especially in the management of the propellant on board, and affect the mission design because side maneuvers now have a much larger (though still small in absolute terms at -0.05 RPM per m/s) effect on the spin rate.

Table 2 shows how the maneuver ΔVs added up for all of the phases.

7 Conclusions

The trajectory design of the ARTEMIS mission that began in July of 2009 has been presented here. The design sent two probes from Earth orbit to the Moon via a transfer that took $\sim$2 years and involved numerous lunar approaches and flybys, low-energy trajectory legs in the Earth-Sun system, and Lissajous orbits around the Earth-Moon Lagrange points on either side of the Moon, and finally culminated with both probes in very eccentric low-lunar orbits. The constraints imposed on the design by the limitations of the THEMIS probes (which were designed for an Earth-orbiting mission)—including thruster orientation, available ΔV, maximum shadow capability, maximum distance for radio telecommunication, and thruster capabilities—necessitated an innovative design. Ultimately the design satisfied all mission constraints and offers a variety of scientific measurement opportunities that have the potential to enhance understanding of Earth-Moon-Sun interactions.

Given the challenges that the ARTEMIS mission presented and the complexity of the design needed to meet those challenges, it is notable that the cost of the mission design effort was many times less than one would estimate for a new, i.e., non-extended, full mission of comparable difficulty. One major difference is that ARTEMIS started in space with given orbits for the two probes, saving the significant cost of determining a launch period and optimal launch targets for the mission. But an even bigger factor in cost savings was acceptance of risk that is unacceptable for a more expensive mission. The THEMIS mission was already a success and completely justified the investment already made in building and launching the probes. Furthermore, the outermost two probes were forced to find a new mission because the THEMIS orbits they were in would have led to fatal shadows by now. So in a sense the only thing at risk was the cost of the ARTEMIS design itself, leading to a situation where the investment at risk was reduced by accepting a higher probability that the risk would be realized.

The primary cost-saving characteristic of the mission design process that put ARTEMIS at risk was the near absence of redundancy, both in the design process and in the products of that process. There is a certain amount of natural redundancy in the use of two probes, and indeed much of the opportunity for new science could be realized even in the absence of one. A significant opportunity would have been missed, though, without the dual measurements that have already been made by the two probes and that are planned for the remainder of the mission. On the ground, however, the design team was pared down so that at times it relied on a single person; had that person been unavailable, a different and uncertain approach for that part of the design would have been required. The limited team size also meant that the design itself was nearly "single string" in the absence of backup and contingency trajectories. The analysis that would have produced such alternative designs was most often replaced by engineering judgment that such alternatives existed and could be found if needed. Similarly, in the area of maneuver design, extensive Monte Carlo runs covering all the ways that reality could diverge from the nominal plan were replaced by experience-based estimates of when trajectory correction maneuvers might be needed and of how much ΔV capability might be needed to correct the trajectories as they were flown.

The greatest uncertainty in the design was perhaps in the area of trans-lunar trajectory corrections because these could contain only minimal ΔV components in the direction of the probe $-Z$ axis. In one of the rare instances of backup analysis, an alternative transfer that included deterministic "down" maneuvers at strategic points along the way was designed; these maneuvers could serve to enable upward corrections by reducing the size of the down maneuvers. But this alternative transfer was not used or needed, and the maneuver design team was able to design TCMs in flight that kept the probes on track to their Lissajous rendezvous. The enabling mitigation of the probe's thrust-direction constraints was that every phase of the mission, including the transfer phase, included multiple orbits of the Earth or Moon so that an up maneuver on one side of the orbit could be replaced by a down maneuver or in some cases a radial maneuver elsewhere in the orbit. Another critical factor of mission success so far has been the stellar performance of the two probes and the mission operations team: every one of the dozens and dozens of maneuvers has been executed as planned.

Acknowledgements The work described in this paper was carried out in part at the Jet Propulsion Laboratory, California Institute of Technology, under a contract with the National Aeronautics and Space Administration.

The authors would like to recognize and compliment the outstanding contributions of the THEMIS/ARTEMIS science team, the ARTEMIS mission design team at the Jet Propulsion Laboratory, the ARTEMIS navigation and maneuver design team at Goddard Space Flight Center, and the THEMIS/ARTEMIS navigation, maneuver design, and operations team at the University of California-Berkeley Space Science Laboratory to the successful development and implementation (so far) of the ARTEMIS mission. Judy Hohl, our editor at UCLA, and Emmanuel Masongsong, our graphics editor at UCLA, contributed significantly to the readability of this paper. The maneuver data in the tables above were supplied by Dan Cosgrove, the THEMIS/ARTEMIS Navigation Lead.

References

V. Angelopoulos, The THEMIS Mission. Space Sci. Rev. **141**, 5–34 (2008). doi:10.1007/s11214-008-9336-1

V. Angelopoulos, The ARTEMIS Mission. Space Sci. Rev. (2010). doi:10.1007/s11214-010-9687-2

V. Angelopoulos, D.G. Sibeck, THEMIS and ARTEMIS. A proposal submitted for the Senior Review 2008 of the Mission Operations and Data Analysis Program for the Heliophysics Operating Missions. Available at http://www.igpp.ucla.edu/public/THEMIS/SCI/Pubs/Proposals%20and%20Reports/HP_SR_2008_THEMIS_SciTech_20080221.pdf (2008)

D. Auslander, J. Cermenska, G. Dalton, M. de laPena, C.K.H. Dharan, W. Donokowski, R. Duck, J. Kim, D. Pankow, A. Plauche, M. Rahmani, S. Sulack, T.F. Tan, P. Turin, T. Williams, Instrument boom mechanisms on the THEMIS satellites; magnetometer, radial wire, and axial booms. Space Sci. Rev. **141**, 185–211 (2008). doi:10.1007/s11214-008-9386-4

J.V. Breakwell, J.V. Brown, The halo family of 3-dimensional periodic orbits in the earth-moon restricted 3-body problem. Celest. Mech. **20**, 389–404 (1979)

S.B. Broschart, M.K. Chung, S.J. Hatch, J.H. Ma, T.H. Sweetser, S.S. Weinstein-Weiss, V. Angelopoulos, Preliminary trajectory design for the Artemis lunar mission, in *AAS/AIAA Astrodynamics Specialists Conference, Pittsburgh, Pennsylvania*, ed. by A.V. Rao, T.A. Lovell, F.K. Chan, L.A. Cangahuala. Advances in the Astronautical Sciences, vol. 134 (Univelt, Inc., San Diego, 2009). American Astronautical Society/American Institute of Aeronautics and Astronautics

S.B. Broschart, T.H. Sweetser, V. Angelopoulos, D.C. Folta, M.A. Woodard, Artemis lunar orbit insertion and science orbit design through 2013. Presented at the 2011 AAS/AIAA Astrodynamics Specialists Meeting, Girdwood, AK, July 31–August 4, 2011, AAS paper 11-509 (2011)

M.K. Chung, V. Angelopoulos, S. Weinstein-Weiss, R. Roncoli, N. Murphy, *Personal email communications, August 13–19* (2005)

R. Farquhar, The utilization of halo orbits in advanced lunar operation. Technical report TN-D6365, NASA, GSFC, Greenbelt, MD, 1971

R.W. Farquhar, A.A. Kamel, Quasi-periodic orbits about the translunar libration point. Celest. Mech. **7**, 458–473 (1973)

D. Folta, T.A. Pavlak, K.C. Howell, M.A. Woodard, D.W. Woodfork, Stationkeeping of Lissajous trajectories in the Earth-Moon system with applications to ARTEMIS, in *Advances in the Astronautical Sciences*, pp. 193–208 (2010)

S. Frey, V. Angelopoulos, M. Bester, J. Bonnell, T. Phan, D. Rummel, Orbit design for the THEMIS mission. Space Sci. Rev. **141**, 61–89 (2008). doi:10.1007/s11214-008-9441-1

G. Gómez, K. Howell, J. Masdemont, C. Simó, Station-keeping strategies for translunar libration point orbits, in *AAS/AIAA Spaceflight Mechanics 1998*, ed. by J. Middour, L. Sackett, L. D'Amario, D. Byrnes. Advances in the Astronautical Sciences, vol. 99 (Univelt, Inc., San Diego, 1998), pp. 949–967

P. Harvey, E. Taylor, R. Sterling, M. Cully, The THEMIS constellation. Space Sci. Rev. **141**, 117–152 (2008). doi:10.1007/s11214-008-9416-2

D. Hoffman, Stationkeeping at the colinear equilibrium points of the earth-moon system, Technical report JSC-26189, NASA (1993)

K.C. Howell, T.M. Keeter, Station-keeping strategies for libration point orbits: target point and floquet mode approaches, in *Proceedings of the AAS/AIAA Spaceflight Mechanics Conference 1995*, ed. by R. Proulx, J. Liu, P. Seidelmann, S. Alfano. Advances in the Astronautical Sciences, vol. 89 (Univelt, Inc., San Diego, 1995), pp. 1377–1396

K.C. Howell, B.T. Barden, M.W. Lo, Application of dynamical systems theory to trajectory design for a libration point mission. J. Astronaut. Sci. **45**(2), 161–178 (1997)

W.S. Koon, M.W. Lo, J.E. Marsden, S.D. Ross, Heteroclinic connections between periodic orbits and resonance transitions in celestial mechanics. Chaos **10**(2), 427–469 (2000). doi:10.1063/1.166509

B.D. Owens, D.P. Cosgrove, J.E. Marchese, J.W. Bonnell, D.H. Pankow, S. Frey, M.G. Bester, Mass ejection anomaly in Lissajous orbit: response and implications for the Artemis mission. Presented at the 2012 AAS/AIAA Spaceflight Mechanics Meeting, Charleston, SC, Jan. 29–Feb. 2, Paper AAS 12-181 (2012)

C. Renault, D. Scheeres, Statistical analysis of control maneuvers in unstable orbital environments. J. Guid. Control Dyn. **26**(5), 758–769 (2003)

D.J. Scheeres, M.D. Guman, B.F. Villac, Stability analysis of planetary satellite orbiters: applications to the Europa orbiter. J. Guid. Control Dyn. **24**(4), 778–787 (2001)

M. Sholl, M. Leeds, J. Holbrook, THEMIS reaction control system—from I&T through early orbit operations, in *Proceedings of the 43rd AIAA/ASME/SAE/ASEE Joint Propulsion Conference & Exhibit*, Cincinnati, OH, USA, 8–11 July 2007 (2007)

G.J. Whiffen, *Static/dynamic for optimizing a useful objective*, United states patent, No. 6,496,741. Issued December 2002 (1999)

G.J. Whiffen, Mystic: implementation of the static dynamic optimal control algorithm for high-fidelity, low-thrust trajectory design, in *Proceedings of the AIAA/ASS Astrodynamics Specialists Conference*, Keystone, Colorado (2006) Paper AIAA 2006-6741

M. Woodard, D. Folta, D. Woodfork, 2009 ARTEMIS: the first mission to the lunar libration points. Presented at the 21st International Symposium on Space Flight Dynamics, Toulouse, France

DOI 10.1007/978-1-4614-9554-3_5
Reprinted from *Space Science Reviews* Journal, DOI 10.1007/s11214-010-9738-8

First Results from ARTEMIS, a New Two-Spacecraft Lunar Mission: Counter-Streaming Plasma Populations in the Lunar Wake

J.S. Halekas · V. Angelopoulos · D.G. Sibeck · K.K. Khurana · C.T. Russell · G.T. Delory · W.M. Farrell · J.P. McFadden · J.W. Bonnell · D. Larson · R.E. Ergun · F. Plaschke · K.H. Glassmeier

Received: 14 July 2010 / Accepted: 14 December 2010 / Published online: 20 January 2011

Abstract We present observations from the first passage through the lunar plasma wake by one of two spacecraft comprising ARTEMIS (Acceleration, Reconnection, Turbulence, and Electrodynamics of the Moon's Interaction with the Sun), a new lunar mission that re-tasks two of five probes from the THEMIS magnetospheric mission. On Feb 13, 2010, ARTEMIS probe P1 passed through the wake at ~3.5 lunar radii downstream from the Moon, in a region between those explored by Wind and the Lunar Prospector, Kaguya, Chandrayaan, and Chang'E missions. ARTEMIS observed interpenetrating proton, alpha particle, and electron populations refilling the wake along magnetic field lines from both flanks. The characteristics of these distributions match expectations from self-similar models of plasma expansion into vacuum, with an asymmetric character likely driven by a combination of a tilted interplanetary magnetic field and an anisotropic incident solar wind electron population. On this flyby, ARTEMIS provided unprecedented measurements of the interpenetrating beams of both electrons and ions naturally produced by the filtration and acceleration effects of electric fields set up during the refilling process. ARTEMIS also measured electrostatic oscillations closely correlated with counter-streaming electron beams in the wake, as previ-

J.S. Halekas (✉) · G.T. Delory · J.P. McFadden · J.W. Bonnell · D. Larson
Space Sciences Laboratory, University of California, Berkeley, CA 94720, USA
e-mail: jazzman@ssl.berkeley.edu

J.S. Halekas · G.T. Delory · W.M. Farrell
NASA's Lunar Science Institute, NASA Ames Research Center, Moffett Field, CA 94035, USA

V. Angelopoulos · K.K. Khurana · C.T. Russell
Institute of Geophysics and Planetary Physics, UCLA, Los Angeles, CA 90095, USA

D.G. Sibeck · W.M. Farrell
NASA Goddard Space Flight Center, Greenbelt, MD 20771, USA

R.E. Ergun
Laboratory for Atmospheric and Space Physics, University of Colorado, Boulder, CO 80303, USA

F. Plaschke · K.H. Glassmeier
Institut für Geophysik und Extraterrestrische Physik, Braunschweig, Germany

ously hypothesized but never before directly measured. These observations demonstrate the capability of the comprehensively instrumented ARTEMIS spacecraft and the potential for new lunar science from this unique two spacecraft constellation.

Keywords Moon · Lunar wake · Counter-streaming distributions

1 Introduction

In contrast to the rather static lunar interior, the Moon's plasma environment is highly dynamic, with the surface fully exposed to the ever-changing solar wind and magnetospheric charged particles. These plasma inputs, which vary by orders of magnitude in velocity, density, and temperature, especially during active solar periods, drive a commensurately variable Moon-plasma interface, which we have studied since the Apollo era (Ness 1972; Schubert and Lichtenstein 1974). Charged particle and photon impacts generate much of the tenuous lunar exosphere and play a role in its loss. They also drive highly variable surface charging on both the sunlit and shadowed hemispheres of the Moon (Freeman et al. 1973; Halekas et al. 2008b). Relatively small and weak, but widely distributed, lunar crustal magnetic anomalies (Dyal et al. 1974) produce perhaps some of the smallest collisionless shocks and magnetospheres in the solar system (Russell and Lichtenstein 1975; Lin et al. 1998; Halekas et al. 2008a; Wieser et al. 2010; Saito et al. 2010). These and other elements of the plasma environment couple to the lunar surface and exosphere, and very likely to lunar dust, in ways not yet completely understood.

The lunar plasma wake exhibits a host of particularly fascinating physical processes. Solar wind ions impact the dayside lunar surface, implanting or scattering, and leaving a void downstream from the Moon. The resulting plasma void has vanishingly low plasma density, and refills only slowly as solar wind ions flow at supersonic speeds past the Moon. The pressure gradient across the wake boundary creates diamagnetic current systems that reduce the magnetic field magnitude in the expansion region near the wake boundary and enhance the field in the central wake (Colburn et al. 1967; Ness et al. 1967; Owen et al. 1996; Halekas et al. 2005). Charged particles refill the wake along magnetic field lines, with the higher thermal speed of the electrons driving a charge-separation electric field that approximately maintains quasi-neutrality while slowing electron expansion and accelerating ions into the wake. Theoretical studies and simulations of this general plasma process have shown that this leads to a rarefaction wave propagating back into the plasma, particle expansion into the vacuum with a velocity related to the ion sound speed, and an exponentially decreasing density on the vacuum side of the interface (Crow et al. 1975; Denavit 1979; Samir et al. 1983). Observations from Wind first demonstrated that this process operates to refill the lunar wake (Ogilvie et al. 1996; Clack et al. 2004), and numerous simulations have since refined our understanding (Farrell et al. 1998; Birch and Chapman 2001a, 2001b; Kallio 2005; Trávnicek and Hellinger 2005; Kimura and Nakagawa 2008). More recently, we have learned that ions can also refill the wake perpendicular to the magnetic field direction, with gyrating protons (Type-I entry) and re-picked up protons scattered from the dayside surface (Type-II entry) both capable of penetrating deep into the wake cavity (Nishino et al. 2009a, 2009b; Holmström et al. 2010). We do not yet fully understand the relative importance of parallel and perpendicular refilling processes in the wake, or how this varies as a function of distance downstream from the Moon.

Many open questions remain about the lunar plasma environment, and each mission to the Moon to date has made new discoveries that have changed our understanding of the Moon-plasma interaction. The latest generation of international lunar missions, namely Kaguya, Chandrayaan, and Chang'E, observed many new phenomena, including magnetic shielding of surface regions (Saito et al. 2010; Wieser et al. 2010), solar wind scattering from the dayside surface (Saito et al. 2008; Wieser et al. 2009), gyrating and re-picked up scattered protons refilling the wake (Nishino et al. 2009a, 2009b), and scattered protons interacting with the wake electric field (Nishino et al. 2009a; Wang et al. 2010). Even now, many components of the lunar plasma environment remain either incompletely understood or unobserved.

Given the opportunity to add to our understanding of this fascinating environment, we retasked two of the probes from the THEMIS (Time History of Events and Macroscale Interactions During Substorms) five-spacecraft magnetospheric constellation mission (Angelopoulos 2008) on an extended mission to the Moon, as a new lunar mission named ARTEMIS (Acceleration, Reconnection, Turbulence, and Electrodynamics of the Moon's Interaction with the Sun). These comprehensively instrumented spacecraft, currently in transition orbits, will eventually enter into elliptical orbits around the Moon, providing unprecedented two-point measurements over a range of altitudes from ~100–10,000 km, with a variety of orbital phasings and probe separations. In this paper, we describe observations from the first lunar flyby through the lunar wake by ARTEMIS P1 (formerly THEMIS-B).

2 Overview of the Feb 13, 2010 ARTEMIS Flyby

The first flyby of the lunar wake by ARTEMIS P1 took place on February 13, 2010. We show observations from several of the ARTEMIS instruments in Fig. 1, and an overview of the geometry of the flyby in Fig. 2. The spacecraft passed through the plasma wake ~3.5 lunar radii ($R_M = 1738$ km) downstream from the Moon, in a region between those observed by Wind and the more recent low-altitude Lunar Prospector, Kaguya, Chang'E, and Chandrayaan missions. The GSE position of the Moon was roughly $[62.2, -8.7, 2.5]$ earth radii, well upstream from the terrestrial magnetosphere and certainly not connected to the terrestrial foreshock, given the large z-component of the magnetic field. The ARTEMIS P1 probe traveled through the wake at an oblique angle, with the orbit trajectory fortuitously aligned nearly along the prevailing magnetic field direction. This afforded us an excellent opportunity to measure parallel wake expansion processes along magnetic field lines. The combination of unfavorable non-perpendicular magnetic field geometry and distance from the Moon apparently precluded us from seeing any effects of Type-I proton gyration entry or Type-II pickup entry perpendicular to field lines (Nishino et al. 2009a, 2009b).

Ion observations by the ESA instrument (McFadden et al. 2008) indicate bi-directional penetration of ions into the wake along magnetic field lines, with roughly exponentially decreasing density in the wake, bounded by an external rarefaction wave, as predicted by models of plasma expansion into vacuum (Samir et al. 1983) and previously observed by Wind (Ogilvie et al. 1996). The characteristic energy loss/gain of the ions on the entry/exit sides of the wake results from ion acceleration along tilted magnetic field lines, which subtracts/adds to the solar wind velocity on the entry/exit sides, as clearly seen in the velocity components. We discuss ion observations further in Sect. 4.

During this first wake flyby, the ESA instrument operated in magnetospheric mode (McFadden et al. 2008), allowing full energy and angular coverage of ions, but with reduced resolution (22.5° angular resolution, 32% fractional energy resolution). In this mode, the instrument can underestimate the magnitude of the solar wind ion density due to the slower energy sweep, and the transverse components of the ion velocity can have small errors because of the reduced angular resolution. Nonetheless, we have compared integrated ion and

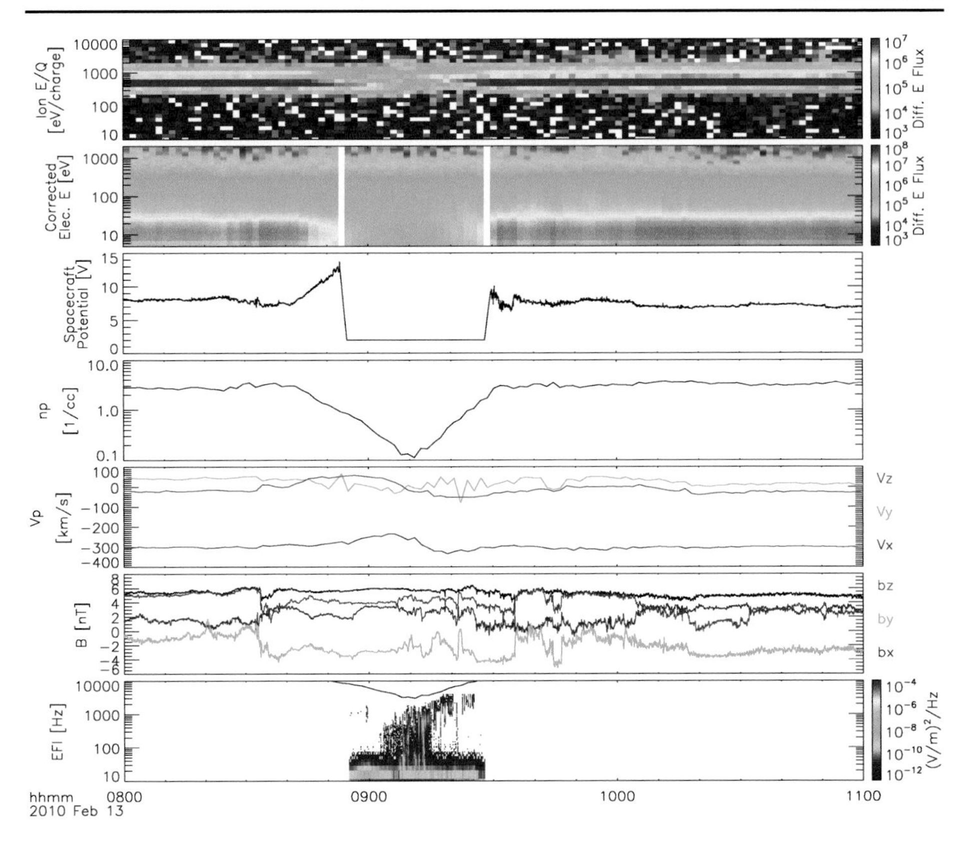

Fig. 1 ARTEMIS observations of the lunar plasma wake on Feb 13, 2010. Panels show omni-directional ion and electron energy-time spectrograms [eV/(cm^2 s sr eV)], spacecraft potential (measured in sunlight, inferred from ion/electron comparisons in shadow), ion density and velocity moments in Selenocentric Solar Ecliptic (SSE) coordinates, magnetic field in SSE coordinates, and an electric field wave power frequency-time spectrogram (*black curve* shows electron plasma frequency). Electron spectra are corrected for the effects of spacecraft charging. Data gaps (*white strips*) indicate times when the spacecraft potential changed too rapidly to apply corrections

electron density and velocity moments outside the wake and found good agreement, indicating accurate measurements of bulk plasma parameters. Future flybys with the ESA instruments in solar wind mode will provide higher resolution measurements of the distributions of solar wind ions as they refill the wake.

In sunlight, the EFI instrument (Bonnell et al. 2008) measures the spacecraft potential (see Fig. 1), allowing us to correct for the effects of spacecraft charging. Just inside the plasma wake (as indicated by the reduction in plasma density, indicating the extent of the rarefaction wave), but outside of the Moon's penumbra, EFI observes an increase in spacecraft potential resulting from the reduction in plasma density. Inside the Moon's penumbra, we tested a new biasing scheme for the EFI probes, with the aim of controlling the spacecraft potential. This biasing scheme successfully moderated spacecraft charging, resulting in a spacecraft potential of approximately two volts positive, as derived from comparison of ion and electron density and velocity moments. This small and well-estimated spacecraft potential allows us to correct for spacecraft charging even in the shadowed wake region, providing a critically important capability for quantitatively analyzing charged particle observations.

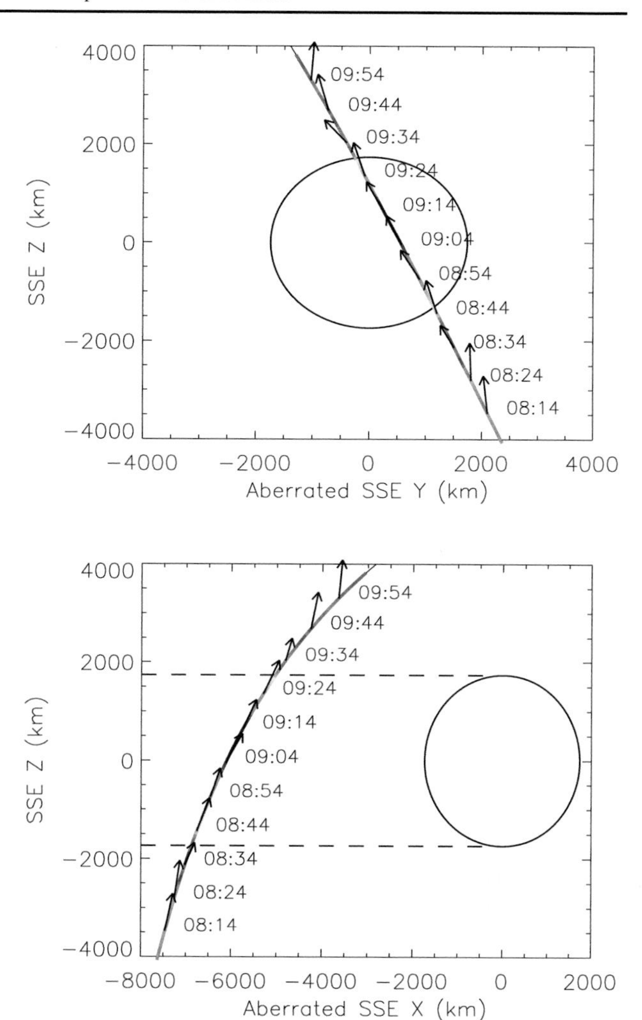

Fig. 2 Geometry of the Feb 13, 2010 ARTEMIS wake flyby, in aberrated SSE coordinates, assuming a perfectly cylindrical lunar wake extended in the direction of the aberrated solar wind flow. We color code the spacecraft track by the log of the ion density, and show magnetic field vectors at the times indicated

To demonstrate this capability, we display electron spectra corrected for the effects of spacecraft charging in Fig. 1, showing a smoothly varying spectrum as a function of time, with no artificial discontinuities at the light/shadow boundaries. Between the penumbra and the umbra, the spacecraft potential changed rapidly, and we did not attempt to correct these electron spectra (blank regions in electron spectrogram in Fig. 1). Electrons, like ions, drop out as the spacecraft travels deeper into the wake, leaving only a residual halo/strahl population in the central wake. We discuss electron observations further in Sect. 5.

EFI also measures the electric field component of plasma waves. During this flyby, as shown in the bottom panel of Fig. 1, EFI observed electric field oscillations extending almost up to the electron plasma frequency, primarily on the exit side of the wake. During this same time period, we observed no magnetic field fluctuations with the SCM instrument (Roux et al. 2008), identifying these oscillations as primarily electrostatic. We discuss possible sources for these waves further in Sect. 5.

The FGM instrument (Auster et al. 2008) measured the vector magnetic field. In shadow, the spacecraft loses the sun pulse, and its spin period drifts slightly due to thermal effects.

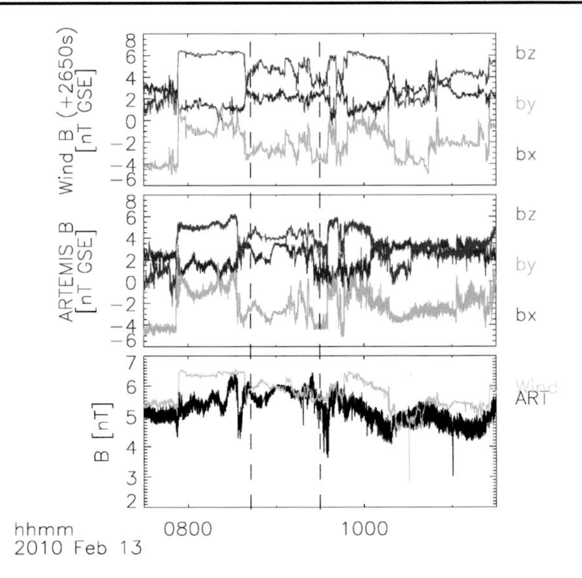

Fig. 3 Magnetic field vector data time series measured by Wind (time-shifted according to the solar wind velocity and spatial separation of the two spacecraft) and ARTEMIS, in GSE coordinates, and a comparison of magnetic field magnitudes measured by the two spacecraft. *Dashed lines* show the extent of the plasma wake as indicated by ion density

The FGM team has utilized magnetic field data, along with modeling of the thermal contraction of the spacecraft, to derive the actual spacecraft spin period in shadow, allowing us to correctly de-spin both magnetic field and plasma data. Magnetic field data thus corrected indicate a highly variable field orientation, but with a prevailing alignment between the magnetic field and the spacecraft trajectory, as shown in Fig. 2. Though we do observe several large rotations in the field, we cannot convincingly associate any of these with the wake boundary. We discuss magnetic field observations in more detail in Sect. 2.

We use the solar wind V_X component measured by ESA, along with the ~30 km/s velocity of the Earth around the Sun, to derive an approximate aberration angle for the solar wind. The V_Y component measured by ESA roughly matches the expected aberration; however, ESA also measures a small V_Z component that we have not utilized in the aberration calculation. As noted above, transverse velocity components measured by ESA can have small errors in magnetospheric mode due to the low angular resolution of the binning scheme. We found that including only the V_Y component provided a better match to the observations, with the ion density dropout that indicates the extent of the plasma wake matching the resulting assumed wake geometry.

3 Magnetic Field Observations

ARTEMIS FGM data indicate a highly variable magnetic field during this time period. In particular, we observe large rotations in the field just before and after the wake transit. One would like to associate these with the wake itself, but we would first have to rule out upstream influences to make such a claim. In Fig. 3, we compare Wind and ARTEMIS magnetic field data. We use the radial separation of the two spacecraft of 133 earth radii and the radial solar wind velocity measured at Wind of 320 km/s to derive an approximate time shift of 2650 s, which we apply to the Wind data. With this time shift, we find that Wind observes essentially all of the magnetic field signatures that ARTEMIS does, with very similar timing. Thus, all of the field rotations have a source far upstream from the Moon.

A comparison of the magnitude of the field between the two spacecraft does yield some interesting differences. For instance, ARTEMIS observes much more significant reductions

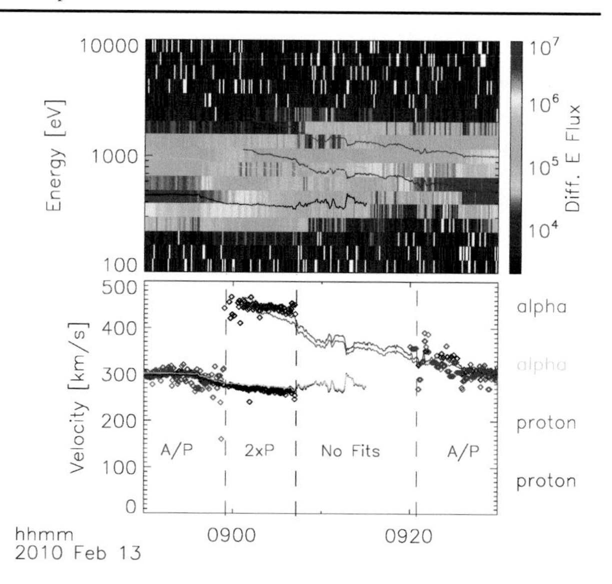

Fig. 4 Ion spectrogram [eV/(cm^2 s sr eV)] (high resolution burst data), and a series of two-component fits to the reduced ion distribution function. "A/P" indicates time periods with distributions best fit to an alpha particle plus a proton component, "2xP" indicates a time period with distributions best fit to two proton components, and "No Fits" indicates a time period with no successful two-component fits. *Red, orange, blue,* and *black curves* on both panels show modeled proton and alpha particle energies and velocities, as labeled in the figure

in field magnitude associated with the large rotations just before and after the wake passage. These magnetic field depressions could represent the expected diamagnetic dips in field magnitude in the expansion region of the wake. However, given their co-location with rotations also observed upstream, we do not favor such an interpretation. To make such a claim, one would have to clearly separate temporal from spatial variations, requiring more than one set of observations, preferably with both observations near the Moon. Furthermore, we find no significant increase in magnetic field magnitude in the central wake, an expected result of the diamagnetic current systems. Therefore, we cannot convincingly associate any magnetic field features seen during this wake passage with a lunar origin. In the coming years, many orbits from two identically instrumented ARTEMIS probes will allow us to sort out features associated with the Moon-plasma interaction from those generated upstream, and separate spatial from temporal variations.

4 Ion Observations and Model Comparisons

As discussed briefly in Sect. 2, ARTEMIS ESA data indicate ions accelerated along magnetic field lines into the wake from both sides. In Fig. 4, we show high-resolution ESA burst data for ions during the wake passage. We also display velocity magnitudes derived from two-component fits (using a gradient-expansion method with a chi square fit metric) to the reduced ion distribution function. This fitting procedure allows a separation of ion species and velocity components difficult to achieve self-consistently in a moment calculation with an instrument like ESA that does not have mass separation. Outside of the wake in the undisturbed solar wind, these fits indicate that proton and alpha particle velocities track very well, as expected. Inside the wake, though, the spectrograms and fits in Fig. 4 indicate the presence of proton and alpha particle components from both sides of the wake, with a complicated spectral structure that we can only fit to a two-component spectrum in a few regions. Wind and other spacecraft have observed accelerated protons like these refilling the wake (Ogilvie et al. 1996), and interpreted them as the result of charge-separation electric fields produced during the expansion process (Samir et al. 1983), but no one to our knowledge has previously observed alpha particle expansion into the lunar wake.

Absent any other information, we would have a difficult time separating the four components of the ion distribution function, composed of interpenetrating protons and alphas from both sides of the wake. However, a model helps us identify the various ion components. Assuming a quasi-neutral 1-D expansion of isothermal plasma into a vacuum, one can derive so-called self-similar expansion relations (Denavit 1979; Samir et al. 1983), which depend only on the combination s/t, where s represents the distance along the magnetic field line into the vacuum, and t the expansion time (or, equivalently, the solar wind convection time past the lunar limb). In all equations, C_s represents the ion sound speed.

$$N = N_0 \exp(-(s/(C_s t) + 1)) \tag{1}$$

$$\phi = -(T_e/e)(s/(C_s t) + 1) \tag{2}$$

$$V_i = s/t + C_s \tag{3}$$

One can extend these relations to take into account non-Maxwellian electron distributions, with the effect of slightly increasing the potential and velocity in the wake. Lunar Prospector observed electron distributions inside and outside the wake at low altitudes, and derived electron properties and wake potentials consistent with this expanded theory (Halekas et al. 2005). Recent observations from Chandrayaan also show that this model may need some modifications near the Moon to take into account surface absorption of plasma (Futaana et al. 2010). For this case, though, we opt to use the most basic self-similar model, demonstrating how well even very simple theory matches ARTEMIS observations.

To develop our model prediction, we use spacecraft position and magnetic field data in aberrated SSE coordinates (as shown in Fig. 2) in order to calculate the convection time t of the point where the magnetic field line intersects the wake boundary, and the distance s along the field line into the wake from that point. Our model depends on the instantaneous magnetic field geometry, and thereby captures the asymmetry of the wake due to the tilted magnetic field. Though one-dimensional in nature, the model implicitly captures two-dimensional features, since convection time maps to distance downstream from the Moon. Our model, utilizing the measured solar wind speed of 310 km/s and an ion sound speed C_s of 25 km/s derived from the measured electron temperature T_e of ~6.5 eV, has no adjustable parameters. Finally, we employ an extension of the self-similar model for alpha particle behavior derived by Singh and Schunk (1982) for polar wind, which assumes that the small fraction of alpha particles do not affect the potential, but rather merely respond to it. Note that this implies that alpha particles sense a different potential than protons, since the two species travel at different velocities into the wake through an explicitly time dependent potential, and therefore we can think of the wake as filtering particles of different species.

We add the model expansion velocity determined for both protons and alpha particles to the initial solar wind velocity in vector fashion to derive the modeled total energies and velocities, which we plot over the data in Fig. 4. Our simple model matches the measured ion energy spectra and two-component velocity fits rather well, correctly reproducing the asymmetry of the wake, and the decrease/increase in energy and velocity on the entry/exit sides of the wake. The model even accurately reproduces numerous small temporal fluctuations (departures from a temporally smoothly varying spectrum) in the ion spectra that result from changes in the magnetic field geometry. This indicates that the infilling ions closely follow the instantaneous magnetic field orientation, as expected given the solar wind convection time of only ~15–20 seconds from the Moon to the spacecraft. With the aid of our model, we can now clearly separate different components of the ion spectrum. We find that protons and alphas from both sides penetrate nearly all the way through the wake, with only the alpha component from the entry side proving difficult to identify in the ion spectrogram. In the

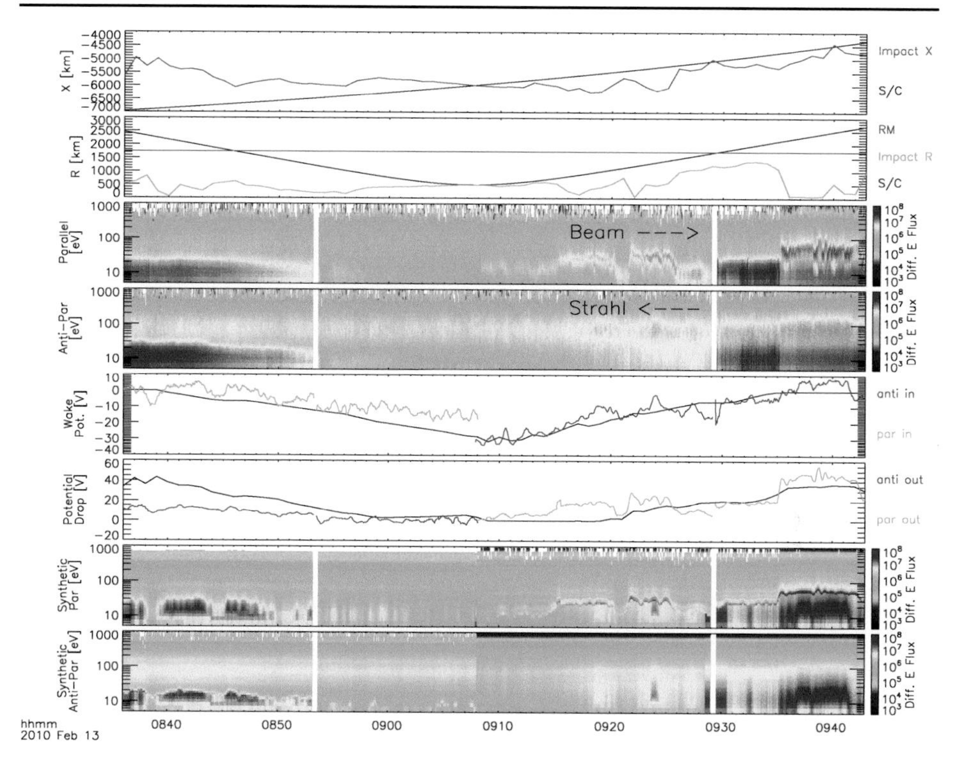

Fig. 5 The *top two panels* show the cylindrical coordinates of the spacecraft and the closest point to the center of the wake reached by the magnetic field line passing through the spacecraft (using a straight-line approximation). The *third* and *fourth panels* show corrected spectrograms [eV/(cm^2 s sr eV)] for parallel (0–15° pitch angle) and anti-parallel (165–180°) traveling electrons (high resolution burst data). The *fifth* and *sixth panels* show the wake potential relative to the solar wind, and the potential drop from the deepest point into the wake along the magnetic field line to the spacecraft, as inferred from shifts in electron distribution functions. *Black curves* on both panels indicate model potentials. The *bottom two panels* show synthetic spectrograms created using the inferred potentials and four end-point electron spectra, as described in the text

central wake, at least three, and probably four, populations of ions coexist simultaneously, each with slightly different vector velocities as a result of the different electric fields felt by the various populations. This promises a rich plasma environment, with a strong likelihood of wave-particle interactions, possibly including beam-beam instabilities as postulated by Farrell et al. (1998). In the coming years, ARTEMIS's full plasma instrumentation and two-probe measurement capability will enable statistical study and clear identification of the effects of interpenetrating ions on the wake environment.

5 Electron Observations and Model Comparisons

As described in Sect. 2, ARTEMIS ESA data show that the bulk of the electron population drops out in the central wake, leaving a residual population sufficient only to balance the small density of interpenetrating refilling ions. The electrons that do penetrate the wake, especially the most field-aligned populations thereof, have fascinating properties. The third and fourth panels of Fig. 5 show two electron spectrograms calculated for parallel and anti-parallel components of the electron population, utilizing high resolution burst data, with both corrected for spacecraft charging as described in Sect. 2. The potential drop set up by the wake refilling process (described in Sects. 3 and 4) excludes most of the lower-energy core electron populations from both entry and exit sides of the wake, though a residual portion of the core population penetrates to the central wake from the entry side. Meanwhile, a strong strahl component travels anti-parallel along the field lines, penetrating all the way through the wake from the exit side and traveling out the entry side. Finally, we find a beam of electrons that travels out from the center of the wake parallel to the field line, and out of the exit side of the wake. ARTEMIS observes this exiting beam of electrons for some time after the wake passage, whenever the magnetic field connects to the central wake.

The top two panels of Fig. 5 show that the energy of this exiting beam correlates well with the proximity with which the magnetic field line at the spacecraft passes near the center of the wake (as estimated by a linear extrapolation of the local magnetic field direction), rather than with the position of the spacecraft itself, suggesting an origin related to the center of the wake. This beam of electrons may therefore have a source in the central wake, with the seed population then undergoing acceleration outwards by the wake electric field. However, the asymmetry of the wake clearly affects the source and/or evolution of the out-going beam, since we do not observe a similar beam on the entry side of the wake.

We can infer the wake potential structure by comparing the field-aligned components of the electron distribution function in the wake with those measured outside the wake, as demonstrated previously using Nozomi and Lunar Prospector data (Futaana et al. 2001; Halekas et al. 2005). We use a parallel electron distribution function measured just outside the entry side of the wake (at 08:36), and an anti-parallel distribution function measured just outside the exit side of the wake (at 09:43), as references to calculate shifts in parallel/anti-parallel electron spectra from the entry/exit sides of the wake to the center of the wake. To accomplish this determination, we use electrons with energies from 40–200 eV (which penetrate all the way through the wake) as tracer populations to derive the potential drop into the wake. This procedure tacitly assumes that the source spectrum of solar wind electrons does not change significantly during the observation time period.

The fifth panel of Fig. 5 shows the results of this derivation, along with a model potential computed from the same self-similar model described in Sect. 4, with exactly the same parameters previously used. The model and data agree rather well, with small fluctuations in the derived potential indicating spatial and/or temporal variations not represented in the model. We find only one major difference from the model, namely an asymmetry between the potential drop derived on the entry and exit sides of the wake. The wake potential may plausibly have a real spatial asymmetry, since the refilling electron populations from the two ends of the flux tube have different properties, and the wake isolates the two ends of the flux tube from each other. The strong strahl population incident on the exit side of the wake very likely affects the plasma expansion, increasing the wake potential drop compared to that on the entry side, particularly deeper in the wake where the strahl dominates the electron density. Alternatively, this asymmetry could result from temporal changes during the wake traverse. Only two-point measurements, which ARTEMIS will provide in coming years, can settle this question conclusively, but we find strong hints that point towards the former interpretation.

By using electron distribution functions from the center of the wake (at 09:08) as references, we can similarly approximately derive the potential drop sensed by electrons traveling outwards from the central wake. This spectral shift depends on the potential drop between the local plasma and the deepest point in the wake on the magnetic field line of the spacecraft, rather than that between the solar wind and the local plasma, with our derivation assuming temporal invariance and a constant electron spectrum throughout the central

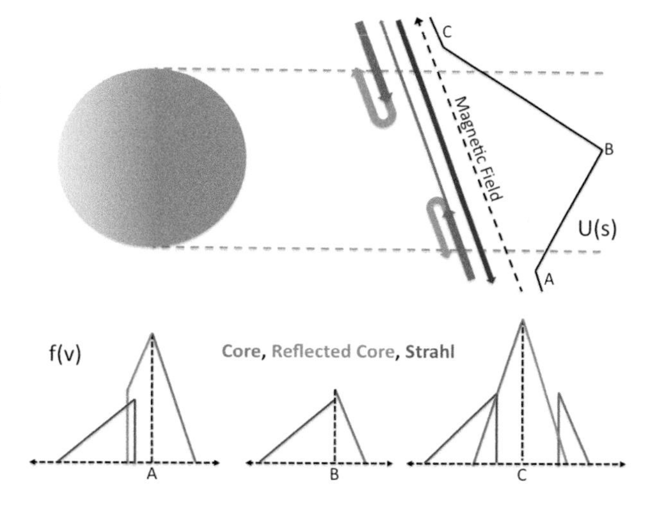

Fig. 6 Schematic illustration of velocity filtration of the electron spectrum by the wake potential, showing distribution functions (at three points) and trajectories of core electrons, reflected core electrons, and strahl electrons

wake. Using these procedures, we can use electron distributions as a probe not only of the local wake potential, but that deeper in the wake along the spacecraft magnetic field line. As above, we compare derived potential drops with the results of our self-similar model, in the sixth panel of Fig. 5. Again, the derived potential drops agree fairly well with our model, except that the model potential drop on the entry side exceeds the actual drop sensed by electrons traveling out from that side of the wake. This implies a potential drop on the entry side of the wake smaller than expected, and smaller than that on the exit side, consistent with the results just discussed. This asymmetry could explain the lack of a clearly observable beam of electrons traveling outward from the entry side of the wake.

As a final check on our inferred potential structure, we derive synthetic spectrograms for parallel and anti-parallel electrons, using the same end-point spectra we used to derive the potentials just discussed. We manufacture these synthetic spectrograms by constructing a superposition of incident electrons, reflected electrons, and electrons accelerated out from the central wake, again using the potential drops derived above to shift the distribution functions appropriately. The resulting superposed spectra match the observed distributions very well, other than small fluctuations likely due to temporal variations. The synthetic spectrograms, based on derived potentials that rely only on measurements above 40 eV, also match the lower energy electrons, with the same basic features as the observed spectrograms. Furthermore, we find that the beam of electrons observed on the exit side of the wake matches very well with the electron spectrum measured in the central wake, shifted by the inferred potential drop from the central wake to the observation point. In other words, this exiting beam most likely consists of a residual population of core/halo electrons that penetrates through the central wake from the entry side and then undergoes acceleration outwards by the wake potential on the exit side. In some regions, this exiting beam co-exists with a reflected population of incident electrons with energies below the beam energy.

In Fig. 6, we show a schematic illustration of the tentatively identified components of the electron spectra in and around the wake, produced by the filtering and acceleration effects of the wake potential. A small portion of the electron core population traveling parallel along the magnetic field from the entry side of the wake penetrates to the central wake, where it feels the electric field set up by the plasma refilling the wake from the exit side, and undergoes acceleration outwards through the wake potential, producing the observed beam. At the same time, few core electrons penetrate from the exit side, but a strong strahl

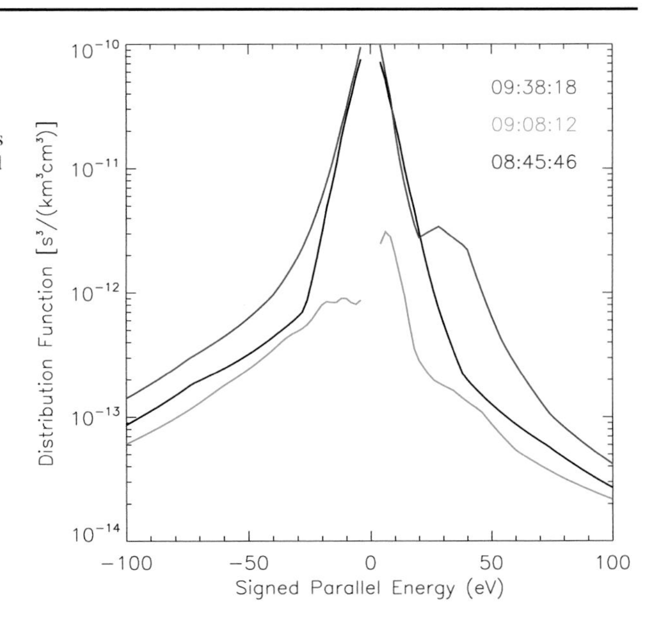

Fig. 7 Parallel cuts through electron distribution functions, covering the same pitch angle ranges as the spectrograms in Fig. 5, as measured at three times before (*black*), during (*blue*), and after (*red*) the wake crossing. Positive values on the horizontal axis represent electrons with velocity parallel to the magnetic field and negative values represent electrons with anti-parallel velocity, for ease of comparison to the schematic spectra in Fig. 6

component travels all the way through the wake and out the other side. In Fig. 7, we show corresponding measured distribution function cuts, demonstrating a striking agreement to those postulated in Fig. 6. The measured spectra naturally show more complexity than the cartoon illustrations, with multiple spectral components (core and halo populations) visible in both the parallel and anti-parallel cuts. However, the key features remain, including the anti-parallel strahl enhancement, the residual parallel core population in the central wake, and the parallel beam formed by the acceleration of this residual population out of the wake. The superposition of these various electron populations plausibly explains all of the features we observe in the electron spectrograms, with no need to postulate any significant temporal variations, though we cannot rule out temporal variations without two-point measurements (which we will soon have with ARTEMIS). A key element of the explanation presented here relies on the asymmetric character of the wake, ensured naturally by the magnetic field line tilt and the different electron distributions incident on the wake from the entry and exit sides. We find it fascinating to speculate on what happens at the boundary between flux tubes connected to the wake and those not connected to the wake, given the possible presence of an asymmetric potential, in addition to differential electron/ion shadowing (Bale et al. 1997), each of which may drive instabilities.

The wake potential first filters incident solar wind electron distributions and then re-accelerates them, naturally producing electron beams exiting the wake. Previous studies have suggested the existence of such filtered distributions, based on observations of wave turbulence outside of the wake (Farrell et al. 1996; Nakagawa et al. 2003). We now have the first direct observations of the electron distributions that could drive such upstream waves. These electron beams can also drive local wave processes in the wake, as can the counter-streaming refilling ion beams. Indeed, EFI measures both broadband electrostatic noise in the central wake, and electrostatic waves near the electron plasma frequency on the exit side of the wake (see Fig. 1). Either ions or electrons may generate the broadband turbulence in the central wake, with previous predictions suggesting a role for the ions (Farrell et al. 1998). The higher frequency oscillations on the exit side of the wake, on the other hand, correlate very well with the exiting electron beam, suggesting that the beam and the oppositely directed strahl electrons generate waves through a streaming instability. We observe these waves at a frequency slightly below the electron plasma frequency, consistent with the downshift expected when beam components comprise a significant fraction of the plasma density (Fuselier et al. 1985). Outside of the wake, core electrons may help damp wave growth, but inside the wake, where the incoming strahl and the exiting beam dominate the electron density, strong wave growth can generate the observed electrostatic oscillations. Birch and Chapman (2001a, 2001b) predicted just such electrostatic oscillations, which we have now observed with ARTEMIS. In the coming years, ARTEMIS will provide many more detailed observations of the wave-particle interactions created by the filtration effects of the wake potential.

6 Conclusions

On the first of many orbits, from one of two ARTEMIS probes, we have already made new and exciting observations of interpenetrating electron, proton, and alpha particle populations refilling the wake from both flanks. The wake potential driven by the refilling process filters and accelerates both ions and electrons, naturally producing unstable distributions that should generate numerous instabilities, including the electrostatic turbulence observed on this orbit. Single-point measurements already suggest the presence of significant asymmetries in the wake, resulting from tilted magnetic fields and anisotropic solar wind electron distributions. Soon, ARTEMIS's two-probe measurement capability will allow us to conclusively separate temporal and spatial effects, and determine the causal flow of many processes hinted at in these observations. ARTEMIS's comprehensive instrumentation and elliptical orbits will also allow investigations of the 3d structure and dynamics of the wake and many other aspects of the lunar plasma environment, including pickup ions from the exosphere and surface, magnetic anomaly interactions, surface electric fields, and no doubt many as yet unanticipated plasma processes. ARTEMIS will also support and complement measurements from other current and planned missions such as LADEE, ILN, and LRO, providing an integral part of the lunar constellation.

Acknowledgements We wish to acknowledge the extraordinary team of scientists and engineers that made the THEMIS/ARTEMIS missions a reality. We thank R. Lepping, K. Ogilvie, and CDAWEB for providing Wind key parameter data. We also acknowledge NASA's Lunar Science Institute for supporting JSH, WMF, and GTD. FP and KHG acknowledge financial support by the German Ministerium für Wirtschaft und Technologie and the Deutsches Zentrum für Luft- und Raumfahrt under grant 50QP0402 is acknowledged.

References

V. Angelopoulos, The THEMIS mission. Space Sci. Rev. **141**, 5–34 (2008)

H.U. Auster, K.H. Glassmeier, W. Magnes, O. Avdogar, W. Baumjohann, D. Constantinescu, D. Fischer, K.H. Fornacon, E. Georgescu, P. Harvey, O. Hillenmaier, R. Kroth, M. Ludlam, Y. Narita, R. Nakamura, K. Okrafka, F. Plaschke, I. Richter, H. Schwarzl, B. Stoll, A. Valavanoglou, M. Wiedemann, The THEMIS fluxgate magnetometer. Space Sci. Rev. **141**, 235–264 (2008)

S.D. Bale, C.J. Owen, J.-L. Bougeret, K. Goetz, P.J. Kellogg, R.P. Lin, R. Manning, S.J. Monson, Evidence of currents and unstable particle distributions in an extended region around the lunar wake. Geophys. Res. Lett. **24**, 1427–1430 (1997)

P.C. Birch, S.C. Chapman, Particle-in-cell simulations of the lunar wake with high phase space resolution. Geophys. Res. Lett. **28**, 219 (2001a)
P.C. Birch, S.C. Chapman, Detailed structure and dynamics in particle-in-cell simulations of the lunar wake. Phys. Plasmas **8**, 4551–4559 (2001b)
J.W. Bonnell, F.S. Mozer, G.T. Delory, A.J. Hull, R.E. Ergun, C.M. Cully, V. Angelopoulos, P.R. Harvey, The electric field instrument (EFI) for THEMIS. Space Sci. Rev. **141**, 303–341 (2008)
D. Clack, J.C. Kasper, A.J. Lazarus, J.T. Steinberg, W.M. Farrell, Wind observations of extreme ion temperature anisotropies in the lunar wake. Geophys. Res. Lett. **31**, L06812 (2004). doi:10.1029/2003GL018298
D.S. Colburn, R.G. Currie, J.D. Mihalov, C.P. Sonett, Diamagnetic solar-wind cavity discovered behind moon. Science **158**, 1040 (1967)
J.E. Crow, P.L. Auer, J.E. Allen, The expansion of plasma into a vacuum. J. Plasma Phys. **14**, 65–76 (1975)
J. Denavit, Collisionless plasma expansion into a vacuum. Phys. Fluids **22**, 1384–1392 (1979)
P. Dyal, C.W. Parkin, W.D. Daily, Magnetism and the interior of the Moon. Rev. Geophys. Space Phys. **12**, 568–591 (1974)
W.M. Farrell, R.J. Fitzenreiter, C.J. Owen, J.B. Byrnes, R.P. Lepping, K.W. Ogilvie, F. Neubauer, Upstream ULF waves and energetic electrons associated with the lunar wake: Detection of precursor activity. Geophys. Res. Lett. **23**, 1271–1274 (1996)
W.M. Farrell, M.L. Kaiser, J.T. Steinberg, S.D. Bale, A simple simulation of a plasma void: Applications to Wind observations of the lunar wake. J. Geophys. Res. **103**, 23653–23660 (1998)
J.W. Freeman Jr., M.A. Fenner, H.K. Hills, Electric potential of the Moon in the solar wind. J. Geophys. Res. **78**, 4560–4567 (1973)
S.A. Fuselier, D.A. Gurnett, R.J. Fitzenreiter, The downshift of electron plasma oscillations in the electron foreshock region. J. Geophys. Res. **90**, 3935–3946 (1985)
Y. Futaana, S. Machida, Y. Saito, A. Matsuoka, H. Hayakawa, Counterstreaming electrons in the near vicinity of the Moon observed by plasma instruments on board NOZOMI. J. Geophys. Res. **106**, 18729–28740 (2001)
Y. Futaana, S. Barabash, M. Weiser, M. Holmstrom, A. Bhardwaj, M.B. Dhanya, R. Sridharan, P. Wurz, A. Schaufelberger, K. Asamura, Protons in the near lunar wake observed by the Sub-keV Atom Reflection Analyzer on board Chandrayaan-1. J. Geophys. Res. **115**, A10248 (2010). doi:10.1029/2010JA015264
J.S. Halekas, S.D. Bale, D.L. Mitchell, R.P. Lin, Magnetic fields and electrons in the lunar plasma wake. J. Geophys. Res. **110**, A07222 (2005). doi:10.1029/2004JA010991
J.S. Halekas, G.T. Delory, D.A. Brain, R.P. Lin, D.L. Mitchell, Density cavity observed over a strong lunar crustal magnetic anomaly in the solar wind: A mini-magnetosphere? Planet. Space Sci. **56/7**, 941–946 (2008a). doi:10.1016/j.pss.2008.01.008
J.S. Halekas, G.T. Delory, R.P. Lin, T.J. Stubbs, W.M. Farrell, Lunar Prospector observations of the electrostatic potential of the lunar surface and its response to incident currents. J. Geophys. Res. **113**, A09102 (2008b). doi:10.1029/2008JA013194
M. Holmström, M. Weiser, S. Barabash, Y. Futaana, A. Bhardwaj, Dynamics of solar wind protons reflected by the Moon. J. Geophys. Res. (2010). doi:10.1029/2009JA014843
E. Kallio, Formation of the lunar wake in quasi-neutral hybrid model. Geophys. Res. Lett. **32**, L06107 (2005). doi:10.1029/2004GL021989
S. Kimura, T. Nakagawa, Electromagnetic full particle simulation of the electric field structure around the moon and the lunar wake. Earth Planets Space **60**, 591–599 (2008)
R.P. Lin, D.L. Mitchell, D.W. Curtis, K.A. Anderson, C.W. Carlson, J. McFadden, M.H. Acuña, L.L. Hood, A. Binder, Lunar surface magnetic fields and their interaction with the solar wind: Results from Lunar Prospector. Science **281**, 1480–1484 (1998)
J.P. McFadden, C.W. Carlson, D. Larson, M. Ludlam, R. Abiad, B. Elliott, P. Turin, M. Marckwordt, V. Angelopoulos, The THEMIS ESA plasma instrument and in-flight calibration. Space Sci. Rev. **141**, 277–302 (2008)
T. Nakagawa, Y. Takahashi, M. Iizima, GEOTAIL observation of upstream ULF waves associated with the lunar wake. Earth Planets Space **55**, 569–580 (2003)
N.F. Ness, Interaction of the solar wind with the Moon, in *Solar Terrestrial Physics/1970, Part II*, ed. by E.R. Dyer (Reidel, Dordrecht, 1972), pp. 159–205
N.F. Ness, K.W. Behannon, C.S. Searce, S.C. Cantarano, Early results from the magnetic field instrument on Lunar Explorer 35. J. Geophys. Res. **72**, 5769–5778 (1967)
M.N. Nishino, K. Maezawa, M. Fujimoto, Y. Saito, S. Yokota, K. Asamura, T. Tanaka, H. Tsunakawa, M. Matsushima, F. Takahashi, T. Terasawa, H. Shibuya, H. Shimizu, Pairwise energy gain-loss feature of solar wind protons in the near-Moon wake. Geophys. Res. Lett. **36**, L12108 (2009a). doi:10.1029/2009GL039049

M.N. Nishino, M. Fujimoto, K. Maezawa, Y. Saito, S. Yokota, K. Asamura, T. Tanaka, H. Tsunakawa, M. Matsushima, F. Takahashi, T. Terasawa, H. Shibuya, H. Shimizu, Solar-wind proton access deep into the near-Moon wake. Geophys. Res. Lett. **36**, L16103 (2009b). doi:10.1029/2009GL039444

K.W. Ogilvie, J.T. Steinberg, R.J. Fitzenreiter, C.J. Owen, A.J. Lazarus, W.M. Farrell, R.B. Torbert, Observations of the lunar plasma wake from the WIND spacecraft on December 27, 1994. Geophys. Res. Lett. **10**, 1255–1258 (1996)

C.J. Owen, R.P. Lepping, K.W. Ogilvie, J.A. Slavin, W.M. Farrell, J.B. Byrnes, The lunar wake at 6.8 R_L: WIND magnetic field observations. Geophys. Res. Lett. **10**, 1263–1266 (1996)

A. Roux, O. le Contel, C. Coillot, A. Bouabdellah, B. de la Porte, D. Alison, S. Ruocco, M.C. Vassal, The search coil magnetometer for THEMIS. Space Sci. Rev. **141**, 265–275 (2008)

C.T. Russell, B.R. Lichtenstein, On the source of lunar limb compression. J. Geophys. Res. **80**, 4700 (1975)

Y. Saito, S. Yokota, T. Tanaka, K. Asamura, M.N. Nishino, M. Fujimoto, H. Tsunakawa, H. Shibuya, M. Matsushima, H. Shimizu, F. Takahashi, T. Mukai, T. Terasawa, Solar wind proton reflection at the lunar surface: Low energy ion measurements by MAP-PACE onboard SELENE (KAGUYA). Geophys. Res. Lett. **35**, L24205 (2008). doi:10.1029/2008GL036077

Y. Saito, S. Yokota, K. Asamura, T. Tanaka, M.N. Nishino, T. Yamamoto, Y. Terakawa, M. Fujimoto, H. Hasegawa, H. Hayakawa, M. Hirahara, M. Hoshino, S. Machida, T. Mukai, T. Nagai, T. Nagatsuma, T. Nakagawa, M. Nakamura, K. Oyama, E. Sagawa, S. Sasaki, K. Seki, I. Shinohara, T. Terasawa, H. Tsunakawa, H. Shibuya, M. Matsushima, H. Shimizu, F. Takahashi, In-flight performance and initial results of Plasma energy Angle and Composition Experiment (PACE) on SELENE (Kaguya). Space Sci. Rev. **154**(1–4), 265–303 (2010). doi:10.1007/s11214-010-9647-x

U. Samir, K.H. Wright Jr., N.H. Stone, The expansion of a plasma into a vacuum: Basic phenomena and processes and applications to space plasma physics. Rev. Geophys. **21**, 1631–1646 (1983)

G. Schubert, B.R. Lichtenstein, Observations of Moon-plasma interactions by orbital and surface experiments. Rev. Geophys. **12**, 592–626 (1974)

N. Singh, R.W. Schunk, Numerical calculations relevant to the initial expansion of the polar wind. J. Geophys. Res. **87**, 9154–9170 (1982)

P. Trávnicek, P. Hellinger, Structure of the lunar wake: Two-dimensional global hybrid simulations. Geophys. Res. Lett. **32**, L06102 (2005). doi:10.1029/2004GL022243

X.-D. Wang, W. Bian, J.-S. Wang, J.-J. Liu, Y.-L. Zou, H.-B. Zhang, C. Lü, J.-Z. Liu, W. Zuo, Y. Su, W.-B. Wen, M. Wang, Z.-Y. Ouyang, C.-L. Li, Acceleration of scattered solar wind protons at the polar terminator of the Moon: Results from Chang'E-1/SWIDs. Geophys. Res. Lett. **37**, L07203 (2010). doi:10.1029/2010GL042891

M. Wieser, S. Barabash, Y. Futaana, M. Holmstrom, A. Bhardwaj, R. Sridharan, M.B. Dhanya, P. Wurz, A. Schaufelberger, K. Asamura, Extremely high reflection of solar wind protons as neutral hydrogen atoms from regolith in space. Planet. Space Sci. (2009). doi:10.1016/j.pss.2009.09.012

M. Wieser, S. Barabash, Y. Futaana, M. Holmstrom, A. Bhardwaj, R. Sridharan, M.B. Dhanya, P. Wurz, A. Schaufelberger, K. Asamura, First observation of a mini-magnetosphere above a lunar magnetic anomaly using energetic neutral atoms. Geophys. Res. Lett. **37**, L015103 (2010). doi:10.1029/2009GL041721

Printed in the United States
By Bookmasters